AF443317

Stations

Listening to the Deep Earth

Stations

Listening to the Deep Earth

Edited by Stuart Hyatt, Janneane Blevins, and W Benjamin Blevins

Stations: Listening to the Deep Earth is the essential listener's companion
to Field Works' *Stations* album. Please find enclosed a
digital download card for the album.

Published by Jap Sam Books

Book and cover design by PRINTtEXT /
Janneane Blevins and W Benjamin Blevins

Contents

We begin *Stations* by acknowledging the traditional Indigenous
inhabitants of the vast lands from which our data emerge.
Scientific research has contributed to extractive economies in
Alaska, past and present. We recognize this ongoing legacy and
commit to honor the cultural knowledge which has served to
steward Alaska's lands past, present, and into the future. *Stations*
brings together multiple voices and polyphonic ways of knowing.
Wherever you find yourself reading this book, take a moment
to look beneath you, listen, offer gratitude, and celebrate the
Indigenous stewards of the ground you occupy.

Prelude

Stuart Hyatt

Ear to the Earth

I'm listening for necessary sounds. Sounds that belong.
Sounds that carry meaning. This is just now occurring to me
after a decade spent recording our planet's sonic wonders
and weaving them into ten albums of music. I'm still trying
to answer a single, fundamental question: *What does the
Earth sound like?*

Stations arose from a long conversation with the Anchorage
Museum, an institution I deeply admire. The museum was
expanding their sound-based exhibitions and programming
in an attempt to reveal and celebrate the natural
soundscapes, music, and spoken word of the region. They
even started an in-house record label: Unbound Records.
When the museum's curators challenged me to engage with
the sonic identities of the Polar North, I was overwhelmed.

Alaska is just, well... enormous. Not just in its physical size
but in the mind-boggling diversity of cultures, landscapes,
and histories. There are dozens of Indigenous languages,
musical styles, and certain industrial and environmental
acoustic conditions that exist nowhere else.

Scouring the museum's online collections, it appeared to me
that the documented sound of human voices and creations
(*anthropophony*), and the audible scope of the animal world
(*biophony*) were already well represented. What I felt might
be underrepresented was fundamental earthly sounds.
We call this *geophony*. It's one category in the framework
of sounds articulated by pioneering soundscape ecologist
Bernie Krause. Indeed, this polar geophony seemed to me
like a necessary sound, one I felt compelled to give more of
my attention to.

So during my first visit to the Anchorage Museum, before
entering the building, I found a small grassy area just
outside the main entrance. I got down on my knees, leaned
forward, turned my head to the side, and placed my ear
to the ground. I closed my eyes and tried listening to the
Earth. I know it seems ridiculous—I attracted curious looks
from museum visitors—but I was singularly focused. How
can I listen to the ground beneath my feet? If I do hear
something, how can I record it?

Considering the audio equipment I already owned, I
hypothesized that I could place a contact microphone on
various surfaces of the Alaskan earth, perhaps comparing
the different sonic conditions floating beneath plains of wet
soil, glacial ice, pine needles, and rock. I could submerge
a condenser mic into the soft dirt and wait for an audible
signal. I could attach recorders to the soles of my feet. I
also knew Alaska was a seismically active region and had
just experienced another round of devastating earthquakes;
perhaps there was something along those lines to record.
But I didn't want to do a project about earthquakes; I was
searching for subtler notes.

I knew the Earth could shout, but could it also whisper?

As I remained awkwardly pressing my ear to the earth, I also realized I didn't just want to sample a single snapshot location of ground-sound. I imagined instead the myriad and often remote places I'd need to record in order to construct some sort of composite sonic Alaska. Based on logistical, financial, and scheduling constraints, this seemed impossible.

So, as I typically do when bewildered by challenging musical ideas, I turned to science. Imagine my utter astonishment and delight when I learned that there just happened to be 280 highly sensitive ground recording devices already in place across every section of the state.

At that moment, *Stations* began.

A New Map

The USArray, funded by the National Science Foundation, is a vast network of several hundred both permanent and transportable seismic monitoring stations spread throughout the United States. Each station consists of a broadband seismometer with associated signal processing, power, and communications equipment, all housed in a severe weather (and polar bear) resistant steel enclosure. We've included portraits of many of these stations throughout the book. They feel to me like lonely little robots, just sitting out there doing their thankless jobs — on the mountains, by the oceans, along streams, in the forests, on the ice — all day, every day, listening to the Earth. The stations have an admirably small environmental footprint, are powered entirely by solar energy, and once removed, leave the ground just as it was before. *(See my interview with seismologist Debi Kilb for a more detailed discussion.)*

In 2014, 280 of these stations were deployed to Alaska and Western Canada, and for the last 7 years, they've beamed real-time seismic and environmental information to central processing computers run by a consortium of government and academic centers. Working in concert, these stations have provided scientists with the granular data needed to map the structure of the Earth with unprecedented detail. This giant experiment — also known as EarthScope — is producing an entirely new cartography of Alaska. But for me, it is a big beautiful recording studio filled with 280 awesome microphones.

What sounds were these stations actually recording, though? Was there a way to find and process the geophony I was so excited to work with? I found a solution in SeisSound, a custom seismic analysis script for Matlab, the ubiquitous scientific computing platform. Amazingly, all the data from the Alaska array is available online in a free searchable database. There's just a lot of it to parse. Using SeisSound, you can dial up any station, scan a timeline of ground movements — however big or small — and export various audible signals from a specific time period. You can endlessly adjust parameters for frequency, sensitivity, duration, location, and axis. The palette of sounds made available through this script are endless. With unlimited audio to render, I found myself with the awkward feeling of having to select (and thus assign value to) one sound or another.

It's a feeling I know well, having invited a group of writers and artists to make sense of the world by assigning value to its sounds for my last collaborative book with PRINTtEXT, *Metaphonics* (Jap Sam Books, 2018). While this assigning of value may have been a fool's errand, it was part of an earnest attempt to move away from a dependence on an image-dominated visual culture and toward a creative practice centered on the simple act of listening.

With *Stations*, I once again became a reluctant arbiter of
meaning, trying to somehow find the *necessary* sounds in
the overwhelming fields of data. I eventually settled into
a groove where I narrowed my focus to a small group of
10 stations with good data logs — representing diverse
geographic locations and topography — and then just let
those individual stations speak for themselves. And speak
they did. This geophony, processed and manipulated by
SeisSound, became the foundation upon which I built the
10 musical tracks for the album. The idea was to then have
a human voice sing along with the Earth's voice in a new
kind of duet. It may feel a bit new age, but it's really quite
beautiful in all its glorious whispers and shouts.

Science and Spirit

My projects invariably end up inhabiting a tiny little space
on a giant Venn diagram where scientific data analysis
and spiritual awakening overlap. In my collaborative series
Field Works, I've relied on the generosity of scientists to
give me access to a rich variety of compositional source
material: a hydrologist sampling the contaminants in a local
stream (*Pogue's Run*), a biologist helping me track down the
endangered Indiana bat (*Ultrasonic*), and now a seismologist
teaching me how to detect subtle movements beneath our
feet (*Stations*).

Field Works even takes its name from the natural sciences,
and my contributions seldom contain the triumphant
moments of inspiration generally associated with studio
creativity. Instead, I trudge along with the meticulous baby
steps of the scientific method. Along the way, a certain
western monotheistic worldview stubbornly persists, and my
role is curiously more akin to that of spiritual seeker than
data scientist. As such, we organized this book (and the
accompanying album) into three sequential sections, loosely
following stages of a transcendent psychological journey:
Invocation, *Reverie*, and *Rhapsody*.

A Fourth Stage

In 2020, a fourth stage emerged: I call it *The Great Quieting*. In March, April, and May, as the pandemic took on its terrifying form, lockdowns spread from community to community. Reports of carless streets and plane-less skies were common, with the associated haunting imagery and dramatically reduced noise.

But what did the Earth itself have to say? The ground, as it turns out, became very quiet. *(See the* Postlude *section for a full seismic quieting report.)* I dug back through the data from the seismic monitoring stations during the lockdown periods, and — spoiler alert — they were indeed calmer. Reduced highway traffic, reduced industrial activity, reduced human *everything*, made the ground become just a little more still. Even in remote locations.

There had to be a fourth chapter to the book and a final track on the album to capture this once-in-a-lifetime sonic phenomenon. It gave me the opportunity to more deeply compare our human bodies with the body of our planet itself. We were both experiencing the same quieting.

Learning Remote Hymns

Western religion's greatest gift to me has been its music, from the majestic European symphonies and operas to the emotionally evocative folk and gospel songs of the American South. Growing up, the only time I ever heard my family or community singing aloud together was at Christmas time. The very act of "caroling" would appear ridiculous in another context or during another time of the year. My own desire to sing aloud in public was slowly shamed out of me over time.

As a young kid, I'd sit cross-legged on the beige carpet in front of my family's turntable, listening repeatedly to Prokofiev's *Peter and the Wolf*. Each character — animals included — had a specific instrument and melody associated with it. I was enchanted, humming along, the entire suite memorized. I remember thinking that the lumbering grandfather *was* the bassoon. The wolf *was* the ominous French horn. I know it's a stretch, but I think this was my first taste of what I can only characterize as animistic music. The flute did not represent the bird — it *was* the bird. Or at least *my* bird.

As I grew older, these things that once provided me such wonder all eventually seemed to get named. These musical animals, I've only recently learned, indeed have a name: leitmotif. It's a technique I've unknowingly inserted into all *Field Works* projects. In *Stations*, the ground itself becomes the central character with a distinct voice. It's now my job to decide which of these voices should be shared, which ones carry meaning, and which ones should be left alone. It's my job to find the necessary music.

One of my favorite recent art books is *Seeing Is Forgetting the Name of the Thing One Sees* (University of California Press, 2018), a collection of conversations between Lawrence Weschler and the American minimalist artist Robert Irwin. The title aptly captures the spirit of Irwin's steady approach to making art, and I am continually seeking an auditory equivalent. Is hearing *forgetting the source of the sound we are listening to*?

A striking passage in the book concerns the "impossibility of a neutral gesture," which in Irwin's case might refer to a mark of paint on a canvas. But taken into consideration when arranging audio field recordings for music, the conundrum remains. I am admittedly not neutral; I am a storyteller with an aesthetic agenda. But when thinking through a potentially animistic field of sounds, I can begin to better approach neutrality as a collector and transmitter

of these very alive voices. I am simply trying to introduce
the songs of our planet to new audiences, and vice versa.
Irwin eventually states that in creating a piece of visual art,
only the "necessary gesture" should command an immediate
presence.

When making audio field recordings, I feel relieved of this
burdensome duty, as it becomes clear that nearly all sources
of biophony and geophony are, in fact, necessary gestures.
They command an immediate presence. They have an inner
power approaching, at least for my purposes, sentience.

Inspirations and Collaborations

I'm indebted to specific precedents and inspirations for
this project. Lotte Geeven's magical recordings made
in a super deep boring hole (30,000 feet!) provided me
endless fascination and encouraged me to listen for
sounds underfoot. And John Luther Adams' renowned
installation *The Place Where You Go to Listen* is an elegant
example of how to generate real-time light and sound from
environmental phenomena (including seismic activity!).

This project has been realized through the tremendous
support from the Anchorage Museum and the National
Geographic Society. I am also indebted to the consortium
of scientists working with seismic data from the Polar
North. However, this book is not really about Alaska at all.
It's about our personal and emotional relationship with an
enchanted ground beneath us, wherever we are. Enrique
Ramirez leads with a whole-Earth sensorium, tracing a
squiggly line through the whirlwind multi-millenium
extraction dance. In *Invocation*, Hanna Benn transcribes a
leitmotif of her vocal duets with the Earth, while Ashon
Crawley carves out space in the landscape for his own
breath, inviting us to breathe with him. Next, Gustavo
Valdivia dispatches a vibrant precolonial tectonic report
from the Andes. In *Reverie*, Paige Lewis unfolds layers

of physics atop pixelated spectrograms. In *Rhapsody*, Joan Naviyuk Kane, Ishmael Angaluuk Hope, and Kashina employ old senses to make new sense of topographies both familiar and strange. In *Quieting*, Anaïs Duplan and Laraaji slow themselves (and us) down alongside a calmer planet. Finally, Debi Kilb brings us to a soft conclusion with hard science.

As you make your way through, I invite you to listen to the necessary sonic gestures that accompany this text. Inside the back cover is a small envelope with a download card to access the entire musical album *Stations*, featuring tracks I created with Hanna Benn, Janie Cowan, Masayoshi Fujita, Laraaji, Qasim Naqvi, and Brad Weber. There are also ten special remixes — peer reviews if you will — by Deantoni Parks, Green-House, Olga Wojciechowska, Afrodeutsche, Nathan Fake, Ben Chatwin, Sophia Loizou, Amulets, Penelope Trappes, and Alva Noto.

I hope you enjoy the music. Feel free to put your ear to the earth and sing along.

Enrique Ramirez

Along the Earth's Sensorium

It was a cool early Spring day in New York. Packets of data, converted into pixelated, low-resolution images of stylized thermometers, suns, and half-moons occluding grids of clouds, and real-time radar sweeps showing bands and chances of precipitation were transmitted through cellular service providers, opened in thousands of apps by multitudes of finger swipes and flicking wrists. A news alert informed thousands of weekend travelers, huddled in underground platforms waiting to travel from Brooklyn to Manhattan, that Northern Italy was in effective shutdown, a consequence of the global outbreak of severe acute respiratory syndrome coronavirus 2 (SARS-CoV-2) that led to an exponential increase of unique views of websites illustrating how to mix aloe gel with rubbing alcohol in order to stave off hand sanitizer shortages in the bodega archipelago. People coughed into sleeves and elbows, mostly their own.

On the 4 train, Lexington Avenue Express, three women talked of the vicissitudes of not wearing parkas while a

busker with dental implants played themes from telenovelas
and a ballad by Selena. Some inside the car began to sing
along, taking pictures and selfies with the same telephones
that previously alerted them that the University of
Washington had suspended classes to mitigate the spread
of Novel Coronavirus Disease 2019, or COVID-19. At the
next stop, a woman and her two kids entered the car and
fell asleep, huddling each other in a mound of puffer jacket
comfort as the subway sped underneath Lexington Avenue.
People exited the platform at 86th Street and ascended into
a field of brilliant white as the light from the sun's noonday
transit bounced off curtain walls and broke up into dappled
patterns along the road surface. Save for a few cirrus bands
scarring the dry air, the skies were clear.

This was only days before the world convulsed in a collective
freakout, before I left the city for the wilds of Hudson
County hoping to ride things out. My first night there, I
was struck by the absolute stillness, a paralyzing quiet that
hung in the air like a shroud. There were no clouds in the
sky, and the moon shone through the skylight above my bed
at a piercing angle, directly in my eyes. I thought of the pink
laser in Philip K. Dick's 1981 novel *VALIS*, how it bestowed
a moment of recognition, a glimpse into the rhythms
of realities otherwise unseen. Molecules conjoin into
nucleotide bands, some forming deoxyribose and ribonucleic
ribbons that weave into globules of data and break apart,
disintegrating into flashes of light that reveal only more
connections. A maze of ridges on a fingertip, rows of
sorghum extending across a coastal plain into an infinite
expanse, a field of burning stars appearing as dust motes
on space telescopes' sensor arrays, a straining glimpse to a
distant fluttering of photons from a galaxy on the haunches
of Ursa Major: all woven into a single reel spooling into
the cerebral mantle before it unravels into the furrows of
its neural pathways. The field of vision narrows to a single,
brilliant point of light, becoming smaller and smaller… and
dilates to reveal memories from my youth.

These were memories of nights when I, too, was lying in my
bed, unable to sleep. I was hearing noises, hissings tuned to

a strange frequency that became a high-pitched yowling, almost like wind. I could not locate these sounds. They were not underneath my bed, and as far as I could tell, there was nothing coming from my brother's bedroom or the hallway outside. I was more curious than scared, wondering how these sounds that I could not locate were present, covering me with curtains of charged particles. One day after science class, I described these noises to my teacher, who told me that they could be one of two things: the reverberation made by blood as it coursed through my circulatory system, or the sounds of electrons colliding with my body. In other words, the noises I was hearing were not coming from an external source. I was only hearing my own body.

The noises that we hear at night, the ones we mistake as being outside of us but are actually inside us — are there moments when the Earth listens to itself in such a way? If so, then the deep guttural rumblings, the scratching and clacking of tectonic plates, and even the crashing of waves, all may be the sounds of human activity, field recordings from a planet with a sensorium as old as time itself. And we can imagine ourselves with something like an atlas of human anatomy, a diagram of the Earth's nervous system presented as a ream of papers. We peel them back from top to bottom, each page flip revealing worlds underneath the Earth's surface. Layer by layer we watch currents of magma spread through the crust, into the mantles, streams that are large at first before they break apart into smaller rivulets, dendrites, and axons that end in fiery blooms far down in the core. And now we reverse course and imagine these smaller streams joining tributaries that flow ever upward until there they begin to flicker. There is a change of state, something hot and smoking becomes electric and cooling, a bluish light touching the mycorrhizal strands that line the undersurfaces of the topsoil, that thread beneath moss, under bogs, and grow into the ice. This is a sensitive Earth, aware, bristling in a constant and delirious babbling with itself. Like the planet Solaris, it is a "protoplasmic ocean brain" lost in a "prodigious and everlasting monologue."[1]

[1] Stanisław Lem, *Solaris*, trans. Joanna Kilmartin and Steve Cox (New York: Harcourt Brace Jovanovich, 1970 [1961]), 22.

We listen to the Earth's sensorium. We hear it pulsing
and cresting along waves, tearing the air apart. We feel it
convulsing underneath our feet. It commits the sounds
of our history to the soil, the rocks, the rings of trees that
when counted backward, conjure a record of changes
imperceptible save these few fleeting traces. The Earth
listens to us, sensing our histories and our testaments from a
world of flesh and blood. These are words from an antiquity
whose forbearers redrew the contours of their surrounding
world. First they were on foot, chasing elusive bands of
sunlight that drove away the cooling earth and revealed a
land ready for digging, tilling, and growing. This was right
before they began scouring the landscape for fleet-footed
creatures that gave energy and warmth in fallow periods.
They slowed down, finally, limiting their movements to
warmer times. They cut down small trees and made light
frames, and over these they stretched skins from their quarry
to protect themselves from the wind and darkening skies.
They came together in clusters in the landscape. Some went
to hillier areas or nestled among the tall grasses. Some made
walls of mud and grass and covered them with bundles of
sticks. Others walked along the rushing waters to those
places where a soil, frothing and loamy, yielded immediately
to stick and spade.

They made quick work of packs of large, hoofed animals
that stalked the grasslands. Some were raised for their meat.
Some were draft beasts that carried their loads and dragged
their blades across the land. Others became caravans that
trotted in lockstep across the steppes, arriving in cities on
watery plains cut with rills that gleamed in the afternoon
sun. The goods were then unloaded onto another kind of
caravan altogether, one made of wooden vessels with prows,
oars, and sails that cast off toward the horizon. At night,
sailors slept on the decks watching stars wheel and turn in
the sky above, leaving fiery trails in their wake. And when
they arrived, it was usually to places appearing as if from
a dream. Places that seemed to have never known of light.
Places with framed tents of mud hollows lit by winnowing
fires. Places where men, women, and children clad in bright
fabrics walked on stone paths between walls that reached

the sky. One of these paths leads from the city, through the rocks, and into an open field. There are lines of men whose polished helmets and spears gleam in the light. They move as one, a giant mass of arms, legs, and weapons, flanked by waves of armored riders kicking a cloud of dust that obscures the sun.

Over rocks, through forests, in and out of saltings and fens, the movement is slow. And with each passing moment, there are changes. The horse drawn chariots increase in size, sticks are bent to sinewy bows, clubs are now quivers filled with arrows. The people change as well. In this semidarkness, bent metal carapaces turn into garish garments riddled with bright buttons. Helmets become feather-topped hats. Spears and maces lengthen into metal tubes to become muzzleloaders and flintlocks. Wicks are struck, burning with gunpowder claimed from earlier exchanges in other lands with caravans and armies bearing bronze lamps and burning with sweet incense.

On through the night they move. And at daybreak, they encounter a large pit in the earth, glowing like another sun. Flames leap out of it, as do gusts of wind that burn their faces and singe their clothes. They cut down trees and build a scaffold. From it, they lower buckets into this flaming maw. In the heat, they shape the glowing masses into long, curving spans that stretch across the earth. They bear carriages that pulse with superheated breathing, churning and driving into the land. They carry more and more people, arriving at dense clusters of homes, of structures that seem to reach ever upward, borne aloft by spikes and rails. The skies are now teeming with things other than birds. Some are round and fat, floating gently in a sauntering breeze. Others creep through the air in a buzzsaw riot, ever increasing in volume, forming giant shapes that leave giant icy scars catching light from a sunrise on the other side of the world.

It is not a sun, but a flash of white light that casts a permanent shadow on ruins from previous wars. In this terror-filled moment, superheated gusts of air rush through cities and landscapes. A great rumbling follows, pulverizing the ruins into mountains of dust and ash. A quiet settles like a blanket, and the world pauses to take a collective breath. In this moment of silence, there are stirrings. Cities grow. They become taller. They spread iron and concrete tendrils into and out of the earth. And high above this, in low orbit, satellites whirl and blaze the electromagnetic spectrum with ceaseless chatter. They are joined by humans adrift in capsules, small lifeboats skipping along the outermost edges of the atmosphere in a perpetual freefall. Some escape to and fling themselves to the moon and back. Some have that rarest of opportunity to glance at their own planet, to witness its clouds whirling across bright blue oceans, to see it hanging in its vast expanse. Some land on the moon and walk on its surface — an incredible moment as they carry rocks and trails of dust that tell their own history of a world forged and committed to a singular orbit, one face pockmarked and scarred, the other forever dark and known only to a handful of lunar probes and satellites. And some of these objects were hurled at impossible velocities, slipping past Mars, the asteroid belt, the Jovian satellites, beyond Jupiter, Saturn, Neptune, Uranus, onward, capturing pictures and sending data to humans huddled in front of brightened panels, eyes bleary and straining as they await the next message from beyond.

These messages are delayed. They hitch rides on high-frequency and infrared bands as they bounce back into the atmosphere. And once they fall to Earth, they are broken apart and converted into small flashes of light. There is a white room, cavernous, filled with the noise of hundreds of technicians in white antistatic garments and vulcanized soles. Their thrumming fingers create a kind of percussive mania, hands flying over keyboards that send these flashes of light along invisible lines that once were paths blazed through forests, then dugways that kept wagons on course through relentless storms. These were replaced by iron veins that scarred the earth, and the hard lines of broken stone and tarred aggregate on which cars sped into the horizon.

And above this, there are white filaments dissolving in the
freezing air, evidence of voyages unseen, travels faster than
the speed of sound, of projectiles flung into a void only to
see the sun rise, and never its setting. The lines linger and
twinkle as they ferry bits of information across this domain,
connecting and joining. The Earth is no longer a collection
of textures, but a firmament, frictionless and expansive.

Eyes open onto a field of flickering colors. They hover,
arranging themselves into a bank of clouds hanging in a
warming sky. And then into leafy canopies that twinkle
in the shifting breeze, sunbeams filtering through a
glimmering screen of white light that ripples like water.
Some are blue-green. Others yellow-red. Alternating zones
of earth and water, like a great map. Everything breaks
apart into pieces, spreading into fields of squares made of
every hue and shade imaginable. Everything is recorded.
Everything is accessible to the Earth's sensorium.

As for those places where our histories intersect with
the Earth's, there are no distinctive features, no defining
attribute save for a single word that invites a deluge of
images: Stations. In antiquity, *stationem* were military
outposts, places where soldiers of any rank stood on two
feet. On any given morning, a *cornicen* slings a brass *cornu*
over his shoulder and plays a pattern of notes over and over,
a kind of coded message heard by the *Magister militium*
on the plains of Agrigentum, who, in his mind's eye,
sees his phalanx arranged there, awaiting an amphibious
assault by the Carthaginians. These sounds hover above the
Mediterranean's watery crossroads, mixing with the clouds
before descending in a deluge that joins the hot air flowing
from the African coast. The supercharged currents move on
an Eastern track, leaving small eddies that drift in clockwise
and counter-clockwise whorls, mixing the dense, cold
waters deep below with the warmer ones above. They run in
graceful arcs, dual circuits of air and water, gyrating through
straits, along the coastlines. They merge with the prevailing
winds that bear the names of older civilizations — *aajej*,
haramata, *ghibli*, *haboob* — traversing through isotherms and
drawing moisture away from the coast. They move in the

air, carrying the voice of a *Megas doux* calling his planners, the *Procuratores ducenarii*, to order. On a crude map of the Eastern basin, they draw lines from where the Bosphorus empties into the waters that carve into the Anatolian plateau, to those crenelated and rocky slips that will protect the praetorian fleet from winds and roving marauders. These are the days when the once-still winds pick up some prattle about a person at the far edges of the Levant, in Judea, Samaria, and Idumea, known as Christus, whose life has become immortalized on the walls of megarons and basilicas with bas reliefs known as "stations."

Stations, written on papers, whispered in hushed tones, and even pronounced as acts or deeds, a word that crossed borders, never to change meaning, always adapting. New, modernized armadas claimed their territorial expanse under this banner of "station." And there was a time when sailing around the world meant going station to station. A brig moored at Nore Station on the banks of the Thames sails to Downs Station on the Kent coast before mooring at Lisbon Station. From there, it crossed the Atlantic and found harbor in Antigua, in the Barbados and Leeward Islands Station, tracking up the coast and spending the winter months first at North America and West Indies Station, then in Newfoundland. And there are other trajectories. Wood-beamed brigs and schooners give way to gallant square-riggers and coal-fired steamers that mark their transits through other ports: Aberdeen to Aultbea Station, then to Jamaica, crossing the line and arriving at South America Station. From there, it is on to Australia Station, then to China Station. The voyage continues northward, through the Bering Strait, in trajectories that mimicked the search for the Northwest Passage from earlier epochs, until an arrival at White Sea Station, near Arkhangelsk, at the westernmost reaches of the Asian continent.

Through a geostationary satellite's mechanized eye, a transpolar flight weaves in the air above the remains of failed expeditions. There are stations here as well. Stations where parked explorers froze to death in search of the magnetic North Pole. Stations where bands of Saami,

Nenet, Khanty, Chukchi, Unangax (Aleut), Yup'ik, Iñupiat, Inuvialiut, and Kalaalit kept food hidden from those animals who arrived in the Spring, starved and prowling. And then there are those stations from which stars were measured, from which the Great Comet of 1811 could be seen in the middle of an arc that began in the southern latitudes and now hung in the skies north of the Arctic Circle, suspended in clouds that refracted the Aurora Borealis. Other stations can be seen from afar, spiked with antennae, and with great radomes and parabolic dishes trained upward. Here, they may catch the traces of numbered signals originating in covert outposts throughout the northern latitudes. They ride on shortwave frequencies, static bursts, and agitated dissonances that congeal into something recognizable. Perhaps it is a human voice — a man's. More than likely it is the sound of a woman and girl reading into a microphone.

Yet the voices we hear are not human. They come from an automated number generator, an analog device that utters numbers punched into a keypad by a wary technician. This too occurs in a station, a ministry office in Warsaw or Szczecin belonging to the Służba Bezpieczeństwa, the Polish Secret Police. They began sending voiced messages like these in the 1950s. So did Mossad, the CIA, and the KGB, clandestine organizations energizing the atmosphere with noise that still hovers above us in ribbons spangled with radiation. To catch them, all we have to do is dial in the appropriate frequency and listen to these traces that were once radio signals, now modulated into digital files encoded in various audio formats. They piggyback on other spectral bursts of information, changing midstream to packets of data, converted into pixelated, low-resolution images of stylized thermometers, suns, and half-moons occluding grids of clouds, and real-time radar sweeps showing bands and chances of precipitation were transmitted through cellular service providers, opened in thousands of apps by multitudes of finger swipes and flicking wrists. Information that here, writing from my station, on this day, I understand that it is an unusually cool spring day in New York.

Invocation

I.

invocation

Hanna Benn

free & gentle, sung like a lullaby

5
v I.
v II.
ah
mmm
mmm
3

9
v I.
ah ah ah ah ah
3
v II.
mmm mmm

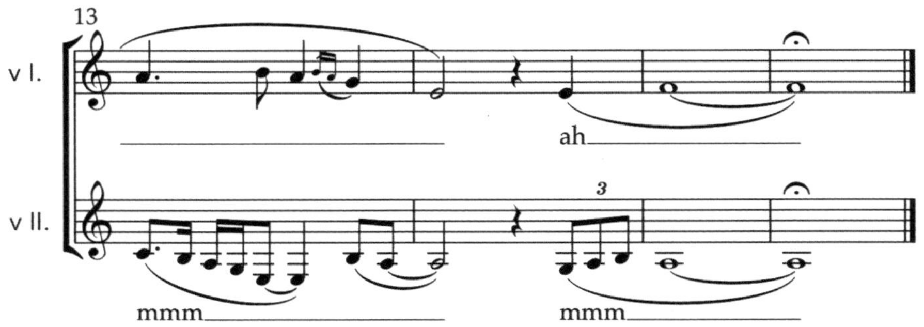
13
v I.
v II.
ah
mmm
mmm
3

Ashon Crawley

preparation.
prayer.

Station 1.

Beautiful. Meaning found when gathering to enter.
To linger in the beauty of entry, of vestibule. To linger
in and wait and tarry in the feeling of announcement. It
encapsulates you, surrounds, but it does not enclose nor
suffocate. It breathes. The porosity allows, lets, yields. Opens.
The acceptance of vulnerability, of fragility.

A flower.

To invoke and convoke and be there in wait and entry
and vestibule.

A prayer.

I…

feel a kind of steady delight, a lightness under and
gathering toward me as the sound vibrates, resonates.
Yes. I feel a yes. Light is the word. Delight is the word.
Luminescence. Shimmer and shine. To begin and enter as
prelude for what is to come. Sense it, sense and feel into it.
It was sent. Sense the sent, the sent for us to sense it. And
it has been sent, sent for us to gather around and withdraw
into the dense field, the folds of its unfolding flower.

The sound of invitation. The sound of welcome. The
sound of rest. As preparation. As what is to come that is not
here but presences through absence. The relation between
tenuous, interrogated through breathed out supplication.
Pulsates. Moves and it moves

us.

Tundra: vast, flat, treeless arctic regions in which the subsoil is permanently frozen. What is a hymnal to tundra, for tundra, in relation to tundra? It is to think about that which is underneath and below, staid and still but still moving, still vibrating. The word, the concept, still: played with, messed with and up and over. Still as rest. But still. Still as continuance and ongoing.

This is a prepositional imposition, the to and for and in, the underneath and below, the with and up and over. A hymn as relation to direction and position and movement.

It is to think about expanse and unending and covering. And wind. To gather the vibrations of tundra and to give them softness and yield and respite. This is song. This, the gift, of *Stations*. Using seismic data from monitoring stations in Alaska and Canada and translating that data into sound. And then translating that sound into song. And then translating that song into the concept of hymn.

Hymns are preparation for worship. They produce occasions for community to sing, to reflect on the cosmic and the material together, to think and sing and emote about gods and creaturely existence. As preparation for worship, they are the enunciations and breaths of adulation and exhortation and praise to the otherwise than earthen vessels. To be vessel, to be in the posture of vessel, to desire to be vessel of earth in which wind and blood and water can move and think and be. To accept earthen things as our existence and condition for gathering. How does one prepare for this?

Stations is preparation for what has already happened, what has already occurred, what has already emerged. The movement of earth and ground, below and beneath the frozen thing. It moves.

Like the breath.

I think a lot about the idea of tundra with relation
to breath. Because I think of temperatures that are near
freezing, if not below, such that the breath would feel sharp
when inhaled and exhaled. Such that my breathing would
be shallow, labored, because of the inhabitation of the
particular landscape.

[space—breathe—space]

But there is also slowness.

[space—breathe—space]

Because the landscape is frozen for much of the year and its low temperatures, everything that grows tarries, waits, has to emerge and flower through and against a landscape that attempts to make such emergence difficult, if not impossible. Yet flowers still grow. And trees still break through the ground surface. Animals breathe and eat and play. This is the practice of living, of black living, of living blackness, a kind of living that emerges against difficulty, if not impossibility. And yet it grows and folds to unfolding flowers. This kind of living, black living, living in the dark, against and through and working with the dark thing.

[space—breathe—space]

Anticipation. Because to wait and exist in wait is to anticipate something otherwise than what is. Is there an ontology for that which remains in anticipatory posture, that which remains in wait? What is the ontology of the thing that is not but expectation, intention, hope? This is not a benign or saccharine or frivolous hope. This is the hope of otherwise than this, a kind of ontology that is against ontology, the being of that which withdraws from that which is because that which is is the attempt towards difficulty, if not impossibility, of life, of living, of blackness.

[space—breathe—breathe—space]

A prayer as contour. To fit within confines and borders
and underneath frozen immovable moving things. The
shape of things to come, the things of expectation, the shape
of desire, of idea, of absence that presences. This is what
the hymn does, it prepares the way, prepares with praise
and adulation and enunciation, that which is hoped for.
Evidence of unseen, but no less real, things.

Space. It is a gap and missing thing that calls out for
indwelling, filling. Tarrying for its own filling up.

This is the music, the hymn that prepares for us a way
to anticipate anticipation, to practice breath as sacred and
moving. And beautiful. To give us chance and opportunity
for rest and respite and hope. Sing together. If not song,
sound together. If not sound, solitude vibrates. Together.

Station 2.

And then, bottom out and feel caressed and held. Be held and behold what it means to feel held. To hold. A kind of enraptured plea and aspiration, the release of breath as the bass note crawls and scrawls and holds. Are you feeling

held?

Harmonic break, like earth, tundra, something over there grows and reciprocates. And repeats. Hearts unfold like flowers, they unfold before us. Held. Not gently but with tenderness. Felt held so you feel the vibration

, and warm.

[hold—breath—held]

Sitting by water at Ivy Creek Natural Area in Charlottesville, I thought about Sally. She sells seashells by the seashore. Such a soft sentence, the repetition of the s sound. A tongue twister, I had to slow down to say it correctly. Things allowed that I wish for, that I want to cultivate: Slowness. Softness. This is prayer and breath and to sing a hymn of slowness and softness is to behold and held to be held and hold together.

[hold—breath—held]

Hold me. Caress me. Like this bass. That bottomed out sound and vibration and movement. The bass moves me. I feel it resonate in me, the hollow within is filled, there is an indwelling the sound makes and makes home in me, finds rest in me. It shimmers in pit, in the essence of my being. Like sitting by water, I

wait.

this is a blackqueer wait unearthed and moving against difficulty and impossibility like breath like breathing in tundra a hymn breathing and still moving and holding my held-in-my-hand heart offering heart to behold being held by some other bass note carry me

here…

Gustavo Valdivia

Vibrant Andes in the Anthropocene

What is transmitted and transmuted throughout this vast evolution is nothing but vibration, vibrations in their specificity, vibrations as they set objects moving in their wake, as they produce harmonic and dissonant vibratory responses. Vibration is the common thread or rhythm running through the universe from its chaotic inorganic interminability to its most intimate forces of inscription on living bodies of all kinds and back again.

Elizabeth Grosz[1]

[1] Grosz, E. A. (2008). *Chaos, territory, art: Deleuze and the framing of the earth.* New York: Columbia University Press.

The Andes mountains, the longest continuous and second-highest continental mountain range on the planet, are the results of a complex geologic history of tectonic pressures, volcanic activity, and climatic interactions. Although this history expands tens to hundreds of millions of years, in the Andes geology is not "buried in the remote past."[2] As these processes are still occurring and causing, for instance, regular — and occasionally catastrophic — seismic activity, human worlds are, as Jane Bennett would say, "inextricably enmeshed" with the vibrant, non-human agencies of the Andean geological forces.[3]

Although it is not clear how pre-Hispanic peoples of the Andes conceived how mountains were formed, or what they considered as the causes of earthquakes, the *Huarochirí Manuscript* — an iconic 16th-century document of Indigenous authorship about the pre-Hispanic Andean religious world — offers a potential clue. In effect, in this text, we are told that *Cuni Raya Vira Cocha*, one of the most important and older divinities in pre-Hispanic Peru, was he "who first *gave shape and force* to the mountains, the forests, the rivers, and all sorts of animals, and to the fields."[4]

The important point to highlight here is that, through this creation act, *Cuni Raya Vira Cocha* not only shaped the topographic features of the Andean landscape but also provided them with force (*kama*).[5] More precisely, he/it ordered, organized, and imbued them with vital force.[6] In this way, as Salomon et al. explain, *kama* circulates reciprocally among human and non-human beings "differently located in action and time," and across the "vital continuum." Therefore, what *Cuni Raya Vira Cocha* ended up establishing was, in fact, a world in which places and people formed part of a larger, interdependent, and animated collective.[7]

[2] Kagan, J (2013) "The not-so-solid earth: remembering new Madrid" In: Kruse, J., & Ellsworth, E. A. (2013). *Making the Geologic Now: Responses to Material Conditions of Contemporary Life*. Brooklyn, NY: Punctum Books.

[3] Bennett, J. (2010). *Vibrant matter: A political ecology of things*. Durham: Duke University Press.

[4] Salomon, F., Urioste, J., & Avila, F. (1991). *The Huarochirí manuscript: A testament of ancient and Colonial Andean religion*. Austin: University of Texas Press.

[5] Cummins, T., & Mannheim, B. (2011). The river around us, the stream within us: The traces of the sun and Inka kinetics. RES: *Anthropology and aesthetics*, 59(1), 5-21.

[6] As Cummins and Mannheim show, in the Andean world "the sun and moon, day and night, and the seasons do not simply exist. Instead, they are *kamachisqa* (ordered, organized, imbued with vital force), a word derived from the root *kama* (to order, to organize, or to have essence)" (íbid. p. 6).

[7] Salomon et. al op. cit.

In pre-Hispanic Peru and also, as various ethnographic studies suggest, in many parts of the Andean world today,[8, 9] people communicate with certain topographic features of the landscape. These landmarks, commonly regarded as *wak'as* in the Quechua language, are not only sacred power-filled places but, principally, persons themselves.[10] In short, as in Salomon's observations about the *Huarochirí Manuscript*, in the Andean world "*wak'as* are made of energized matter, like everything else, and they act within nature, not over and outside it as Western supernaturals do."[11]

[8] Gose, P. (1994). *Deathly waters and hungry mountains: Agrarian ritual and class formation in an Andean town*. Toronto: University of Toronto.

[9] Ricard, L. X., (2007). *Ladrones de sombra: El universo religioso de los pastores de Ausangate (Andes surperuanos)*. Cuzco and Lima: Centro de Estudios Regionales Andinos Bartolome de las Casas (Cuzco), & Instituto Frances de Estudios Andinos.

[10] Allen, C. J. (2002). *The hold life has: Coca and cultural identity in an Andean community*. Washington, D.C: Smithsonian Institution Press.

[11] Salomon et. al op. cit. p. 19.

[12] Salomon et. al, *The Huarochirí manuscript*.

[13] Pachacamac had a majestic sanctuary near modern Lima, a vast ceremonial center that maintained its prominence as a religious and ritual settlement for over 1,000 years until the Spanish conquest in the 16th century.

[14] Swenson. E. and Jennings J. (2018) "Introduction: Place, Landscape, and Power in the Ancient Andes." In Jennings, J., & In Swenson, E. (2018). *Powerful places in the ancient Andes*. Albuquerque: University of New Mexico Press.

As the *Huarochirí Manuscript* describes, even Tupac Inka Yupanqui — one of the most powerful and feared Inka rulers — was a generous venerator of many *wak'as* located in different parts of the Inka territory. The Inka frequently served them with lavish offerings which, as we know today, not only included "gold, silver, clothes, food" but also, almost "everything [he] possessed." [12]

Among all the *wa'kas*, Tupac Inka Yupanqui had a special relation with Pachacamac, one of the two most powerful *wak'as* in all pre-Hispanic Peru. Pachacamac,[13] which translates from Quechua as "the force that animates and sustains *pacha*, the earth/space-time"[14] — was also known as

Pacha Cuyuchic, the World Shaker. In effect, this great Andean deity was not only revered as a wise and influential oracle but also feared for his devastating power to shake the earth as an expression of his anger. As the *Huarochirí Manuscript* describes him/it: "[When Pachacamac] gets angry, earth trembles. When he turns his face sideways it quakes. Lest that happen he holds his face still. The world would end if he ever rolled over."[15]

For over twelve years, we read in the *Huarochirí Manuscript*, the Inka's army had been fighting, unsuccessfully, to suppress a group of rebel communities. In this context, Tupac Inka Yupanqui decided to call a group of *wa'kas* together in the city of Cuzco, "[f]rom every single village … who have received gold or silver,"[16] to consult them on military matters. The Inka was worried and downhearted, as the numbers of his army were being seriously diminished by the rebel forces. Once the *wak'as* arrived, the Inka explained the unfortunate situation of his army, and pleaded with them, in return for his constant offerings, to intervene in his military adventures to stop his enemies.

[15] Salomon et. al op. cit. p. 113.

[16] Salomon et. al op. cit. p. 114.

After listening to the Inka, all the *wak'as* remained silent.
This unexpected situation irritated the Inka and decided to
compel the *wak'as* to accept his petition: "Shall the people
you've made and fostered perish in this way, savaging one
another?" And then he continued: "Why should I serve you
and adorn you with my gold, with my silver, with basketfuls
of my foods and drinks, with my llamas and everything else
I have? Now that you've heard the greatness of my grief,
won't you come to my aid?" Finally, at the peak of his anger,
the Inka threatened all the *wak'as* that were listening to him
in Cuzco: "If you do refuse, you'll burn immediately."[17] It
was only at that point that Pachacamac spoke up: "As for
me, I didn't reply because I am a power who would shake
you and the whole world around you. It wouldn't be those
enemies alone whom I would destroy, but you as well. And
the entire world would end with you. That's why I've sat
silent."[18]

How should we understand this legendary passage from
the *Huarochirí Manuscript*? Furthermore, what kind of
synesthetic language sustained this relation and made both
the *wak'as'* and the Inka's erratic personalities affectively
comprehensible to each other, or, as Goodman would say,
"accessible, audible, and thinkable through words"?[19] What
does it entail that, among all the *wak'as*, it is the great *wak'a*
Pachacamac — the only one associated with earthquakes —
who decides to speak up first? What did it mean to the Inka
that, after he threatened to burn the *wa'kas*, Pachacamac
revealed that his destructive/creative extra-human power
was so overwhelmingly great that if it were ever to manifest
itself fully, would destroy all humans?

[17] Salomon et. al, *The Huarochirí manuscript*, p. 115.

[18] Salomon et. al ibid.

[19] Goodman, S., & MIT Press. (2012). *Sonic warfare: Sound, affect, and the ecology of fear*. Cambridge: The MIT Press

While these questions may go unanswered, one thing is certain: encounters between humans and *wak'as* were never fully predictable in the Andes. On the contrary, it seems that these were instances that could, as Jane Bennett would say, "aid or destroy, enrich or disable, ennoble or degrade" whoever engaged in them.[20] By revealing the entanglements between the social and cosmic domains, the contingency of these encounters were significant occasions in which relationships of mutual obligation and reciprocity between the human and non-human worlds came into being.

[20] Bennett, *Vibrant Matter*, p. ix.

In some way, this is probably something that Alexander von Humboldt anticipated after his own encounter with the Andean geological forces, many years after the incident between Pachacamac and Tupac Inka Yupanqui, as he experienced the Cumana earthquake while he was travelling in South America in 1799. As we know from his account, this earthquake left a "depth and peculiar impression" in his mind, and allowed him to discover the transforming power of dynamism as a quality of our planet. He wrote:

> *When [we] suddenly feel the ground move beneath us, a mysterious and natural force with which we are previously unacquainted is revealed to us as an active disturbance of stability. A moment destroys the illusion of a whole life; our deceptive faith in the repose of nature vanishes, and we feel transported, as it were, into a realm of unknown destructive forces. Every sound — the faintest motion in the air — arrests our attention, and we no longer trust the ground on which we stand.*[21]

What kind of power do these geological and "unknown destructive" forces have to produce such radical transformations in our subjective worlds and in our ways of knowing the world? As Douglas Kahn reminds us, even if earthquakes sometimes are not heard, as their ripples circulate, moving the ground "from one end of the earth through to another," they can always be *felt*.[22]

[21] Humboldt, A. (1848). Cosmos, *a sketch of a physical description of the Universe*. Translated from the German by E.C. Otté. London: H.G. Bohn. P. 215

[22] Kahn, D. (2013). Earth sound earth signal: Energies and earth magnitude in the arts. Berkeley: University of California Press. p, 137.

[23] Yusoff, K. (2018). The Anthropocene and geographies of geopower. In *Handbook on the geographies of power*. Edward Elgar Publishing.

[24] Bennett, J. (2015) "The Shapes of Odradek and the Edges of Perception." In: Textures of the Anthropocene: Grain, Vapor, Ray. Vol. 3: Vapor. Ed. Katrin Klingan et al. Cambridge: MIT Press, 2015. 13-28.

[25] Voegelin, S. (2021). *Sonic Possible Worlds, Revised Edition: Hearing the Continuum of Sound*. New York: Bloomsbury. p. 158.

If my non-orthodox reading of the *Huarochirí Manuscript* is correct, the passage I presented reveals that for a long time, in the Andes, "geophysical events have been in constant material communication with life."[23] In this way, rendering ourselves sensitive to the vibration of Earth forces that, although inaudible to the human ear — as when the *wak'as* did not respond to the Inka — or even "too vague or sharp… too fast or slow … too smooth or intermittent for the human sensorium,"[24] can bring other shades and fundamental elements to produce a more sophisticated account of the interconnectedness that characterizes the geohistorical present of the Andes. Furthermore, bringing sensory attentiveness to nonhuman forces "that we cannot yet hear"[25] could be a powerful way to challenge dominant understandings of the Andean world, especially those that are situated and shaped by the ideological constructions of colonial and Peruvian state forces, which continue transforming the Andean ecologies in the context of the Anthropocene.

Reverie

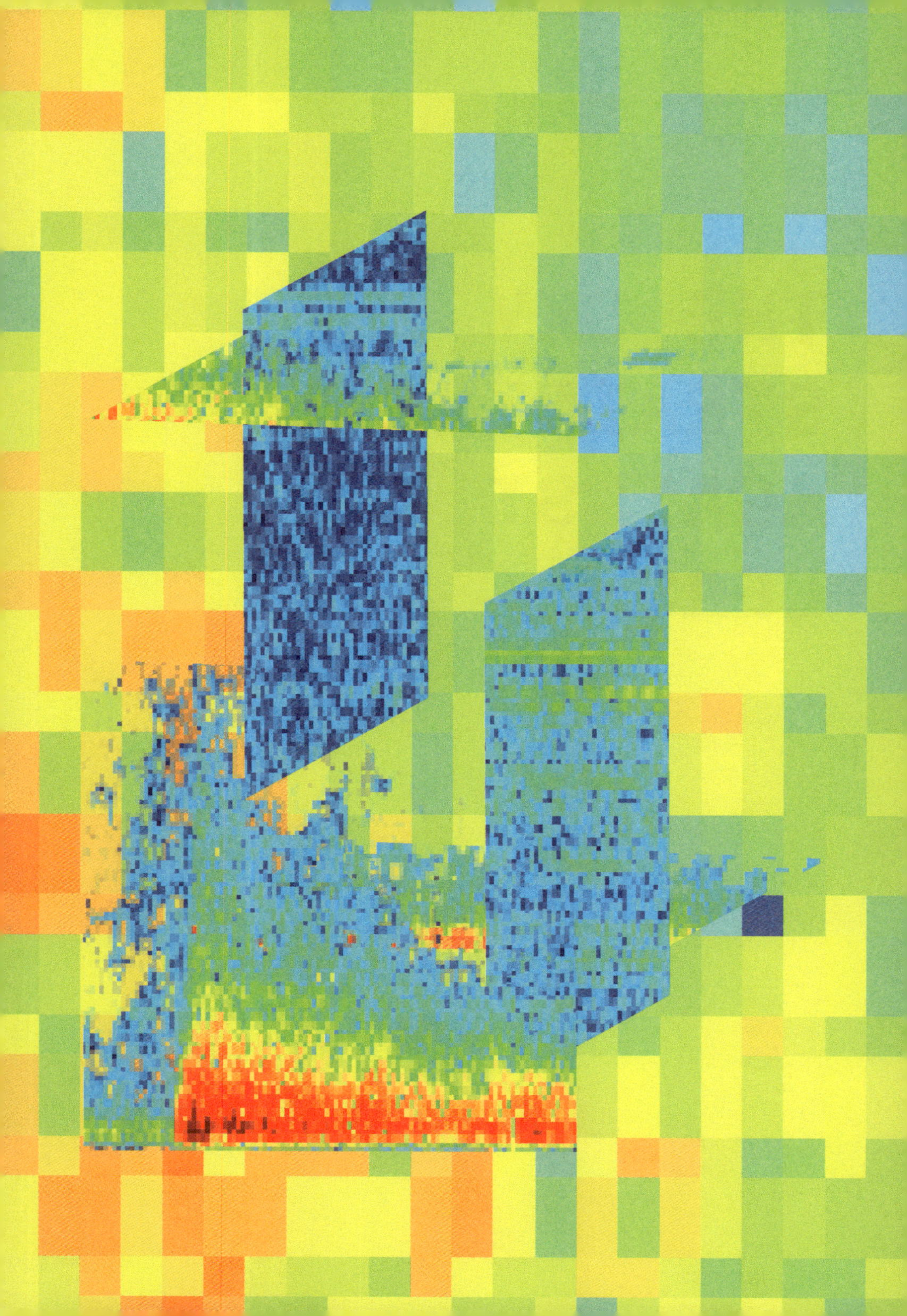

Paige Lewis

THE SKY IS STRAINING AGAINST ITS LEASH

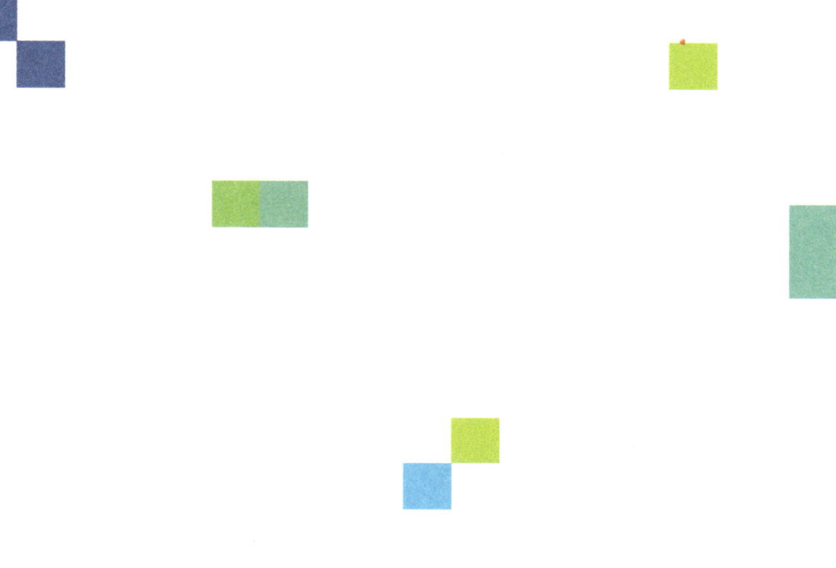

I've got the soul of Bruegel haunting up my pillowcase.
There's no other explanation for why I wake every morning

with a new masterpiece creased across my face. Yesterday,
a crosshatching of fish eating other fish. Today, a horse,

a plow, and Icarus drowning under my right eye. He's
so lonely it makes me want to climb out of my skin

and drive to the beach. I stop at the farmer's market
where the man with a perpetually bruised fingernail

sells bread that tastes like soap to me, but it's so expensive
that it must be good, so I buy it and bring it to the shore.

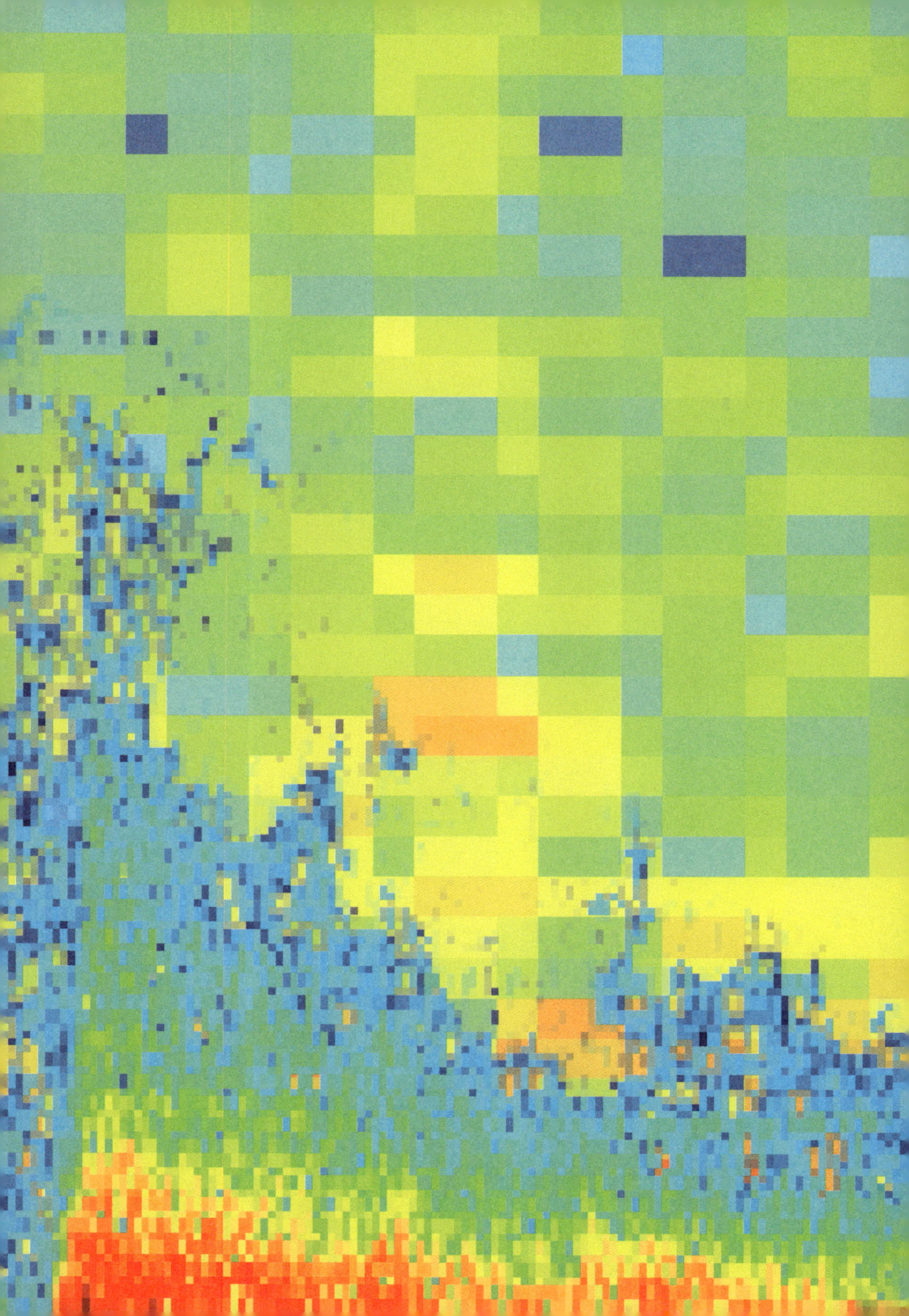

I only bite what I don't understand, which means I've
been biting everything lately: the daily paper, roller rinks

and here, this cacophonous ocean. I fold it up into the size
of a lozenge and pop it into my mouth. The lifeguard begs

me to spit it out. You can't just do that here. You have to be
mathematical and honest with gravity. Gravity has certainly

taken an interest in me—I've become the heaviest corner
of the earth. I've abandoned tallness, deemed too dangerous

for my blood. In the 1900s, at least a dozen men scaled
buildings for sport. What they had in courage, they lacked

in originality, each naming themselves The Human Fly.
The city below them was filled with people and the sound

of people. The birds there were forced to sing at night. My
favorite Human Fly was the man who would intentionally

slip from one balcony to another. A practiced falling to build
suspense. I'm convinced balconies were invented so that

women could go outside at night without being harmed. I
have no balcony and I am certain I have seen the moon

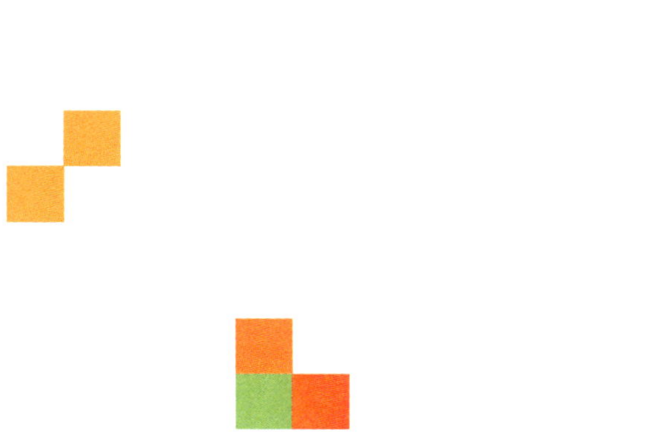

more often in photos. Though on this beach, I am able
to look beyond myself, and I look out over the sea, focusing

on the curved horizon until it becomes not horizon. Becomes
a line between two blues. I reach out, press my palm against

the line. Solid surface. Damp. When I pull my hand
back—Oh, I've messed it up, left a big handprint, muddying

the horizon. Sea blue leaks into sky blue. Could this be a form
of dissection? The earliest dissections were done to find

where the soul resided. Even I know that the soul isn't
physical. If it was it would be all sharp right angles. We
wouldn't need to search for it, we'd see it peeking through
our soft shoulders, transcendent as a Grecian mirror. Ancient

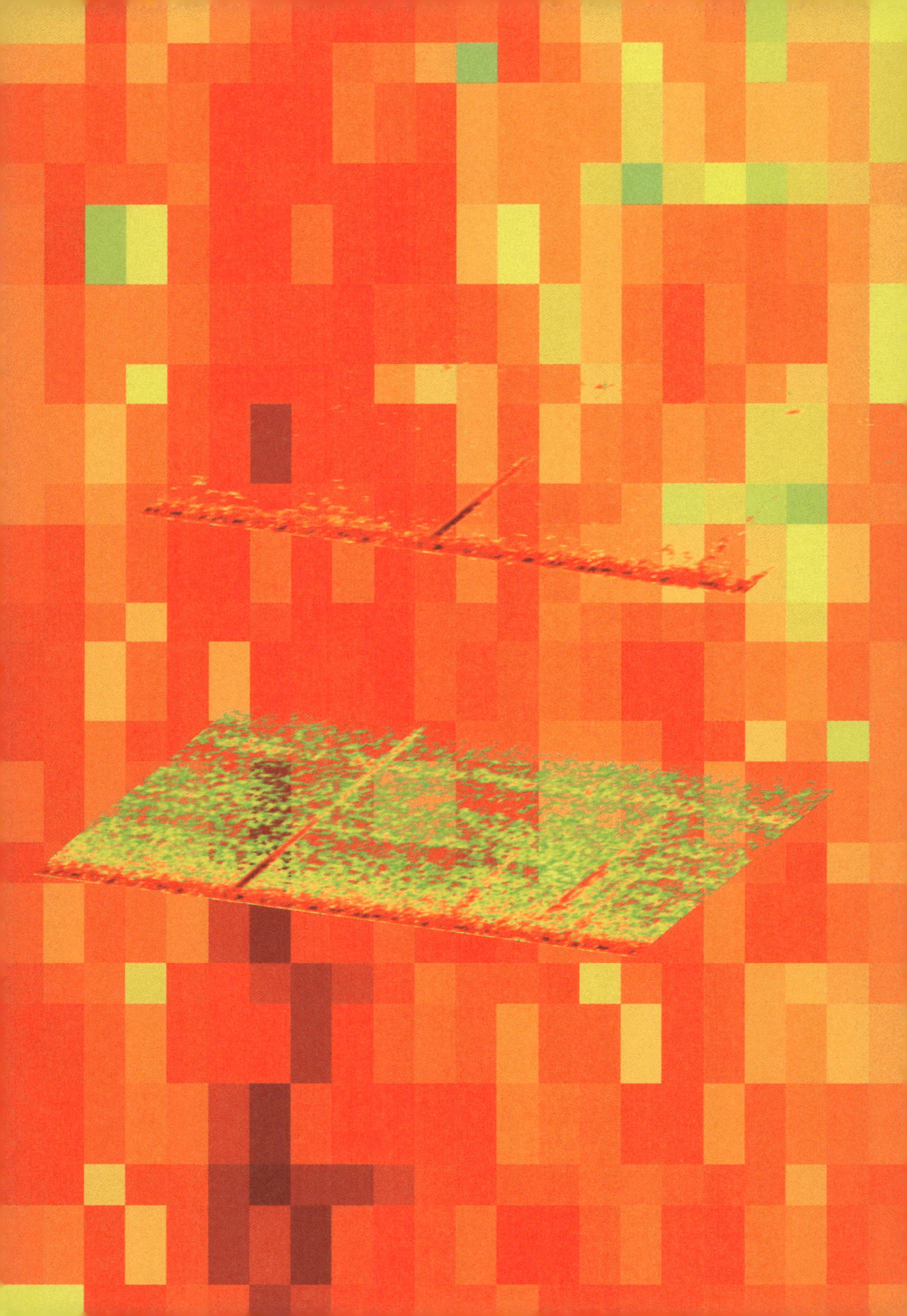

Greeks used to decide which bits of language to prioritize
by laying their words out in the grass and surrounding them

with a ring of fire. Whichever words were saved first were
deemed the most important. I was the first word saved.

Forgive my on and on— my blood feels especially racy
today. I want to say so much I start mixing up the facts,

claiming that a broken bird will transform into two new
birds, that the earth has a heartbeat comparable to my own.

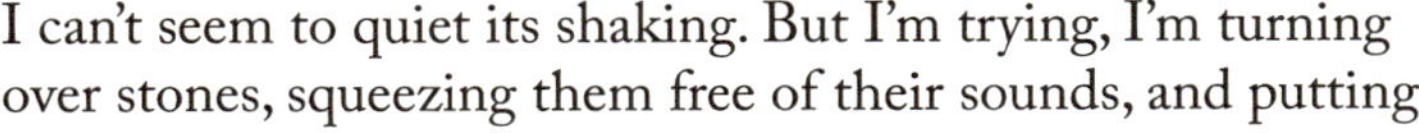

I can't seem to quiet its shaking. But I'm trying, I'm turning
over stones, squeezing them free of their sounds, and putting

the speechless ones in my pockets. I always empty my pockets
before going to bed with someone. In the morning, I will

leave the stones behind on the bedside table. A new shrine
to the quiet person I've become.

Rhapsody

Joan Naviyuk Kane

onomasticon

onomasticon

suppose a bluescoured sky

    though I found no hook, no
anchor, nothing to hold me—

               & wrote
       from the bright, cold spring
& there he led us in circles

in   a   box
    in   a   city
       in a   minaret

                    opponent

                         I became nothing more
                than something to storm
        scale & then subtract

        O
to have been the penultimate she let
she left—syrinx-- to be cut to rock
or something visible to all

Kashina

Magnifications

Kashina was sent packets and packages of organic matter from across Alaska, about which she wrote a series of sensory-vignettes.

Cattail — Chansh Kaq' Bena

Bristling sausage woolbind of soft. Skewered. One tap means no, two means come here. I'll come, resounding taps on ground, rippled fat under the surface, faint cream dimple underneath the felt, closed eyes in profile. Woolboll waxing. An explosion of tiny matchsticks with preburnt tips, wheat and beaver. Silver woodshave crescents. Pulled off in masses, cereal tawn rug to stretch myself over and bury into. Ball unfurl intimacy. Underbelly's secret. I press, you pulse back. Your outer coat smells gently resinous, interior fawn and silent, you float upwards and cling to my lips when I bend near. Sleepsome. Abundance confined, a density of weavings, wish-travels.

Photographs by Kashina

Rosehips — Chansh Kaq' Bena

Playful dehydrated fruit goji sweet crab-spider-small-apple acid in compote form, a tea a chew a toy — longarmed starfish wreath-centred. Sticky buds extruding a slow smile lip's drag of spittle, interiorly sweet. Permanent rain drops. Child's garden of malleable textures as crown atop, squeeze and tang, roll-me-up into squidge of vermilion, bulbous vessel, jammed. Below tiny thorns that damage quickly, hook into fingers, leave skin with dents and shreds, is all the blossoming and sugar a sorrying for their unruly starts? Needles spright men's cologne sportly, the clearest bracing spark. Scales, husking, grasses, take it all in with my hands, they stay with me, magnetized.

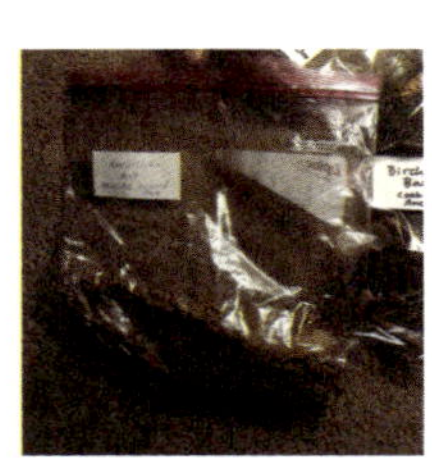

Devil's club and dirt — Ułchena bada Huch'iłyut

Sheer evil. Passive mound of dirt, incubating amidst the soft flour, demerara sugar, greyish sand-cast and moulding material . . . branch encased with sharp needles, relentless barbs, instant-injure. Poisonous caterpillar. Nonstop, all the way down embedding and prickling. Obscene hand gestures. Golem melting to mass of tacks and pins. The dirt surrounding you has a slight sparkle, trapped tears glinting, quartz and glycerine eye winking, shine from an art deco brooch lost in the grains — dry, fine, this could be sand on a tropical beach, instead you harbour and ensnare, massed scorn incarnate, corolla of spikes, thorned outward growing horror.

Birch tree bark — Dena'ina Ełnena

Reverse a straggle of rough black bark, flashback to the astragalus used in medicinal soups — heavy, weighted, already charred, coalblack. Turned around its melt arrested, snow greys, dusty Band-Aid pink, patterned bark (three Norns in hoods, papier-mâché chew, stalactites, monasteries built in mountains, boiled wool and unloosening felt, bird blood stain, beetles with long horns climbing upwards), celadon lichen attached to the scramble, star-splotch and tiny translucent copper buds dangling, cedarheart rose'd, thin and malleable, damp rich scent of water-dripping caves — olive shoots. Silver hairsystem of frizzy roots lively and uncoiling, curved scroll, moon landing layers eyes/mouths/ knees + gashes hidden throughout, a SPEAKING BARK.

Indian celery aka pushke aka cow parsnip —
Dena'ina Ełnena

Baby's rattle, witch's broom holding seedlet pompoms aloft,
small and ridged like fennel or caraway — they collect in
a heap, some insectine with two miniscule antennae at
one end, others shaped like okra or goat's horn peppers.
Minor protuberances crawling up stalk as if they wanted
to become blownaway seedlets too. Choose me choose me.
You look like you should be crushed in a mortar and pestle,
you'd be monk's seasoning, for chastening — austere scent
of chrysanthemum, mouldering floor boards, dark cells,
shadows. In dried form nothing bright, ants on a log Cel-
Ray'd or root parsnip sweet — it's cemetery gloom, last rites
aspergillum-umbel.

Mayday Tree — Dena'ina Ełnena

Bird's eye pimento berry juiced to bursting on the stem. Tannic goblet. Gently serrated leaves, warted tan, open-palmed leathery-skinned. Rub between the fingers yields plum dark running and more pit than pulp. Inky, need a filter to separate the grit from the prune, teastained soil, sandy and moist. Indigo enamel. Enough small particles to file your teeth down. A twig of red beads, percussive, bell-smell, sonorous, festival spray. Checkerboard red and black. I jump over yours. Twig declaiming. Vocal and active branches. Outstretched slenderings like candlesticks, push pins, golf tees. Stick the pins in where you have been in the universe.

Leaves — Tikatahtnu - Deggech' Dgheyay' Kaq'

Mille-feuille crackle, crisp underfoot, make your boot the filling for these rolled-out wafers, a variety pack, parchment of warm shades, dulled penny, salad bowl, sheep snout, meat crackling . . . fragile, thin, and desiccated, their stems still retain a lively pig tail's curl, swing a left hook, J for Jump and L for Lick, they still want to clasp and grab at all surfaces, hold onto my arm darling for support through the treacherous blizzard, barrel full of simians hanging on to the icicled roof and rapping their tails against the windowpane slyly, staring at your food — scent of secluded vaults, latex bandages, silicone dolly-mould.

Labrador Tea — multiple shrubs — Qizhjeh Vena

Commas eyebrows swings — upward-growing peapod
— cradles lined with a dusty tan. Crinkly avocado-
green skin, when you comb through them you feel a dry
electric energy, sleeping animal. Inside the boats there's
no slumber but a harsh, bitter, scrubby smell, overturned
ink, blackened splotch-spill, this'll cure ya, vats boiling,
unfriendly and stayaway like burnt photographs,
explosion in the darkroom/developing lab . . . False lead-
on of rosemary and pine that runs away quickly, leaving
me with this boulder. Strong: to-wrap-my-arms-around,
learning to harness, how do I control the inkblots with
mental power, steering the marks with your mind . . .

Hybrid B/W spruce tips, single tree — Qizhjeh Vena

Dynamic force-field-filled steadying world spinning off in miniature, no-one could tell you were part of a larger growth, snipping did nothing to temper you. It remains aloof, hardy, already-adult, ready to instruct (stand up straight brush your teeth and shoes), I listen I listen I listen to this upright no-nonsense tree springing up perfect straightaway, all the needles single and independent, durable everlast scent that mirrors its form — no softness or slouch — cooling analgesic mountainous-clearing spring-invigorating.

My lungs fill, I shed, I lighten. The pale green pith when you strip the needles, faintly sticky and armour-spiky, the needles hold onto the brown skin like tails, sprouts, rockets, more growth leads to growth leads to growth neverend.

Branches — Tikatahtnu - Deggech' Dgheyay' Kaq'

Eelskin and shark sleek, dotted with constellation-stars,
paler markings, walking and crawling from season to season,
curved and slinky lines revealing flax blond hair serpentine
when peeled, vegetal platinum kiss to feed on, sweetening
and accelerating in barley sugar, tasty and distil-worthy,
spirituous, calorific . . . "Barbara the Fair with the Silken
Hair." The buds greet you with hands to shake, soft and
pawed, with a scent to match, fresh and vivacious, well-
groomed and tended-to, a pleasant disposition. A long,
burred protuberance interrupts the tangle, coarse and brushy,
raccoon peering in, the scullery maid left a trace a yarn-
tangle of her daylight life by accident at the dream ball.

Snow in jar (picked up 3/8/21) — Tikatahtnu - Deggech' Dgheyay' Kaq'

Hiding/cloudy/saline/like Pocari Sweat/has suspended white particles of whey/rock/shell/bone china/algae/sea fleas. Brown bud-shells float swim and sink in the brine. Liquid cloud. Mini-ocean. Living aquarium. Puncture holes in the top and let the larvae breathe. I never got to see it in frozen form ready to put maple syrup on (fresh-fallen) hop-on or -over piles by the sidewalk, throw play sigh at ... it's gone through many invisible secret transformations, which one is revealed now? Sweet snow fay gone, no windkissed cheeks or mittening here — instead a swirl of soft bloomy Morbier washed rind, penicillum Roqueforti spores, fromagerie/charcuterie discarded skins macerated in salt from sobsoaked handkerchief, stagnant pool and boiled brassicas, break the coconut open and it spills perspiration and glandular secretions over the floor. Our cow has been bewitched and gives clabbered milk.

Dirt — Tikatahtnu - Deggech' Dgheyay' Kaq'

Gardener's earth — powdery masses of tea and straw that
fall apart when pressed, easy give and disintegration, ridged
twig bits, bramble, string remnant, feather, reed or fish-
bone and empty ink refills, the smallest seagull dissection,
flour settling into fingerprints, leaf matter, spice mix, pale
infant pebbles that may grow into large crystal mountains =
difficult to picture these as separate entities as together they
create a bristling anthill of chatter — they march, converse,
and huddle — sweepings, filings in the Etch-a-Sketch,
blown away, salt pellets. When flattened with the palm, the
moss webs together like a hanging basket — you can shape
the soil into lozenges, faces, it's a flexible canvas, universal
glue, shifting, adaptable, dare-it-dare-it it can do all.

Shrubs and grasses — Mamterilleq

Lacemade soft and candlewax pale faded trim on antique
dresses found in attic chest . . . dainty florettes like coriander
seeds, ghost lanterns or peppercorns, ready to be behind-
glass and framed or embroidered. Stem ridged and papery,
pages of an uncut book. Newsprint-ink scent stark and grey.
Long oatcake grass waiting to be sheaved, idly chewed or
tucked behind the ear. Curled-up tobacco-brown leaves
rolling in on themselves, crossed arms, hiding dying moulder
of damp stairwell and condensation on concrete. A palmful
of doll-sized sprays, a palmful of fox-colour stems: shining
buds that break apart into wet squeeze shards, cocoa-
chartreuse bright as spiky smiling mascots on chestnut-
cookie boxes, but smell like a nose-pinching squiggle of
honey mustard — look down into the alder cones — hint of
rosy insides and a soapy poppy star when cut.

Wild blueberries — "picked just North of the Alaska Range. Around 64° Latitude."

Ambrosia! Hurrahing at the deep violent-ultraviolet juice — the joy of running into, how impoverished and staid EVERYTHING is compared to the crazed fuchsia flow of these, popsicle stand, liquid light displays, squishes of tomato festals, wine stomps, proudly displaying your berrystained up-to-ankles . . . Doesn't this make you feel more active? Don't you want eight mouths and ten hands to strip a bush clean? Hands henna-tipped nail-lacquered and braceleted with juice. A good summer squeeze rolling the little brighter-than garnet-amethyst in the fingers, party-popper jamsomecookiefillinglinzerheartglassesfruitrollupbythemilewaterslidestreamer geysersfermentforth . . .

ofpruneraisinetteexuberancehahafallingfreelyendlessblueskyrichnessdiviningrodgush

Cottonwood buds, single tree — Qizhjeh Vena

Taffy pulled. Soft glowing points covered in resin, dip of asparagus spear in treacle, choco-lime lively — sugar-crisps bundled up, friendly conglomeration of auburn sap on top of below sideways all together — wrest the wrapper from the sweet — citrus-inflected warming smoke and honeydrip — curative and spirit-lifter, joy that such generosity could exist — laughing as scent tumbles cartwheels claps out of the pen-nibs, paintbrushes, beaks — rootbeer soda fountain, a float with two straws and fat-speckled jellybeans on the side, playful invitation for anisic chew — singly an intimate kissed smell, tinted with acid, open-mouthed, bodily — the bud splays open between the fingers spongily, it gushes yellowed resin, surprise molten centre, leaves ochre stains on the fingers, residual goldenshine. Heavy with all their nesting doll gifts.

Ishmael Angaluuk Hope

DISCOURSE PATTERNS

The breath
organizes thought,
rearranges and makes strange,
bounds over its own mind, seeks glitches,
like Elders mapping the land
as if words have material body.
With each brushstroke,
I gesture to the abstract expressionists,
and the surrealists, too,
who were exposed to the old truths,
the totem carvers and mask makers,
burning from within,
like Rilke's vision of Apollo's torso,
like bioluminescent fur.
They ride the world's knife edge,
slip glimpses, make sculptures
so real they blink,
set off tiny love bombs in the heart,
tell more than what they know.
Stories as nutrient-rich
as watershed eddies,
nourishing the salmon
who settle in the cold.

Quieting

Anaïs Duplan

Tundra Hymnal

It's about being quiet I know a thing or two
about that to speak   even in this, I need a protocol
                                        to illuminate to straighten up to reap quiet
                                        I can do what I've been asked to
settle the sadness   I occupy a room look for how I fill out space   with apology
not a step longer, not a moment longer   the slowing rhythm, which approaches seismograph
holds it in its hand   laughing I can see her
                              across ice in her dress a small breeze sits
                                                            in her there are
                                                gentler tones, her Bakhtin,
                                                which I had read earlier
in the day gave him a choice   as to whether he wanted to be real
                                        when you fall out but when in flow
                                        you see there is flow
                                                      so when I breathe
                                                      from the mouth
                                                      and not the nose,
                                                      I feel I have an impact

I am a part of what's happened here   I'm prone   in-between the palm of my hand I,
sing softly for some gentleman   to come collect me
                                             you see I worry about feeling engulfed

does anyone else
            witness haunt inhabit believe   in tangents this choir   part of my soul

                                                      I heard it
                                                across from the station it was
                                                mastered I could live out
                        from this place I mean, say, I mean my fingers taken hold
            some piece of meaning   at odds with
                                                manifest here   I can here
words before they come out
I can here this
manufactures a beat and starts to
amount to what   I'll try to do is hear the whole thing   see what I can see
                                                            I love this voice
                                                            how it returns ah oh
                                                      shoo be doo bee I want to be
                                                      free Fred Moten said it turns
            out he knows this accumulation
            of noise the question

of why seems a simple stupid question   also totally unanswerable

this hole swallow up what falls
into it and feels full up   I take a walk heard him say
                    what kinds of folks will have me   what exactly have I been
          asked to do here speak
                    I'm already   I run out of hurt my little heart would you please just
          adapt to me and I'll try coming out to meet you
          outside yourself come on over I promise to cry about
          and to be on guard the earth is quiet   because people are inside
                                                on their computers and is being a song
                                                I remember that particular night another
                                                series of nights exactly the same surging
                                                I mean if you felt sick why not drop out
                                                the earth?
                                                            fetch for your pains let it run
                                                            this pendulum back and forth
                                                            you know it's like trying to
          make sense car crash
          say let go of the wheel   you'll be alright I think OK
                                    maybe how much pain that requires breaking yourself
                                    and let's be honest open each and every day
                                    to see the earth is music is relational
                                                            the emotions I graph or do I
                                                meant graft onto the world
                                                I'm learning to trust my face wonder
                                                about surveillance, the police state
                                                when at my computer the language
                                                and go and meet together in a dark
                                                parking lot to talk about it more

never really wanted more than   I can also trust my arms fingers
                                    if that extends to my brain but my face
                                                and arms my fingers my speech
                                                I need to tell you this chorus been
                                                singing to me the earth is quiet
because people are inside on their beds writing kind of like a miracle
not as hot not as cold
I feel laughter at the end
could only be sinister   but joy you can't mistake it that kind of feeling,
                    which I know other people feel   because I've felt and seen
          them I want to be warm

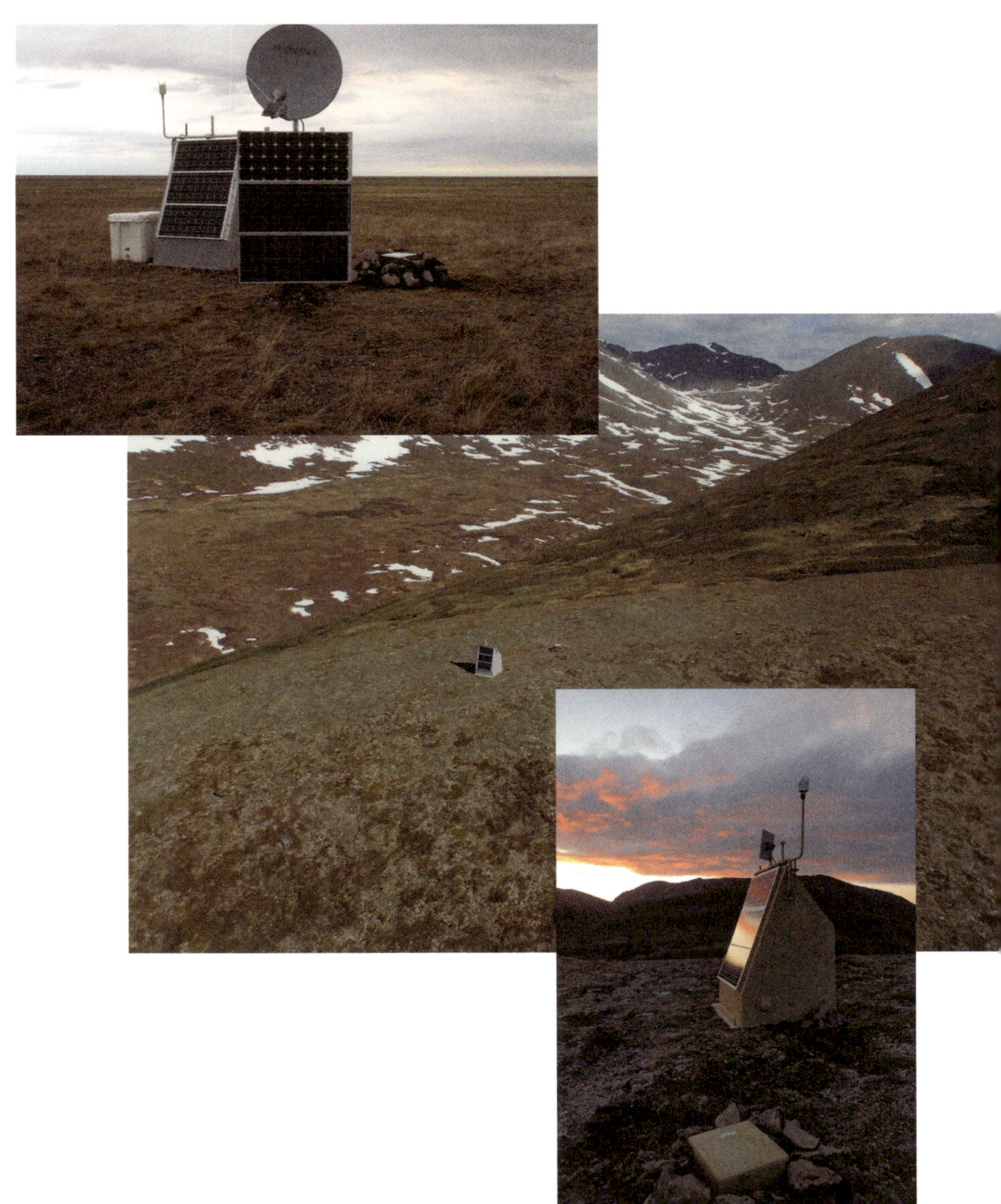

Laraaji

THE
QUIETING

Stills from a film by Hanna Craig.

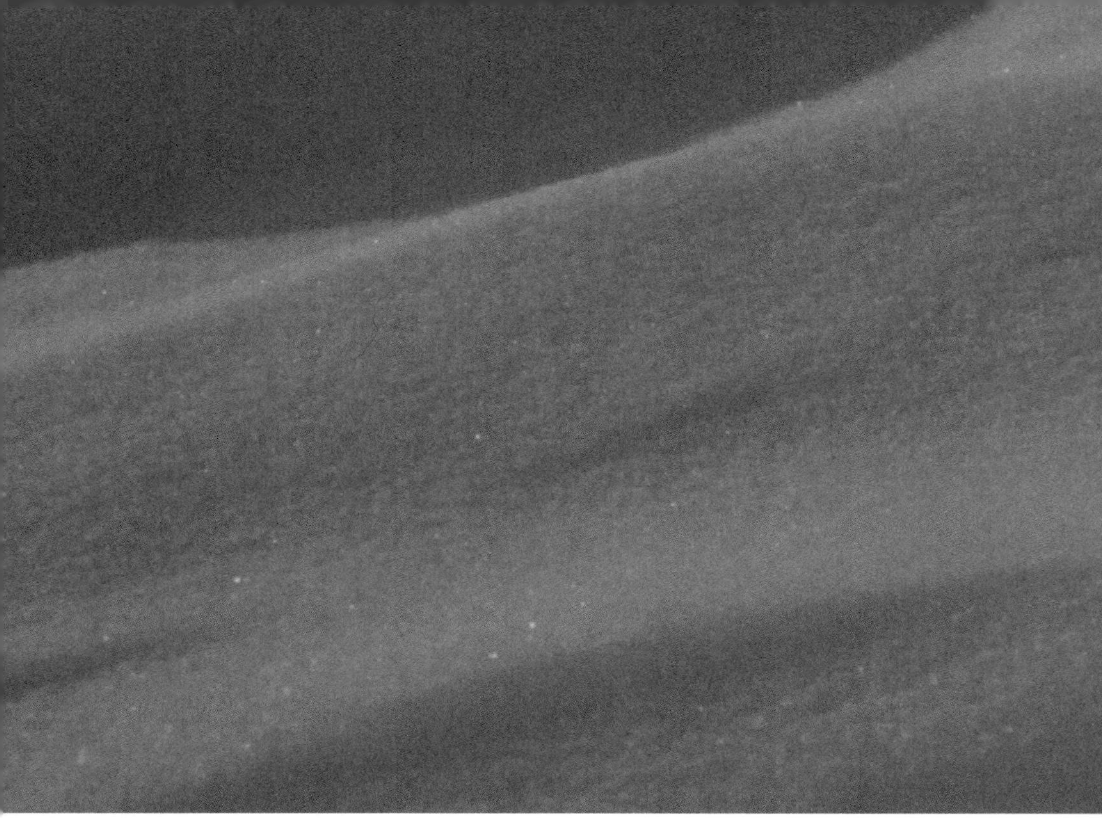

The quieting is
the bringing of awareness
onto an awe-
some plane
of inner self memory

where natural stillness dominates
the mind and body are being rescued from an old and weary normal.
The quieting could be called a "great chill"
accompanied, for the prepared ones, by rapture and spiritual bliss.
a great chilling of external activity.

like a sudden inner expansion
of mass holographic self memory.
the language of the quieting is quiet
itself,
it is silence itself,

it is the mind with decreased external entanglement,
it is the heart with decreased external provocation,
it is the body with dramatically altered outer agenda.
A great lightening up.

the quieting is an immersion of our human species in
well-monitored distancing and isolation,
and the gifting of our beautiful species with extended space
and time for deep inner self witnessing.
The human impact on planetary environment is shifting.

Public gathering activity is quieter,
and for personal space time flow
simplicity is the new convenience.
Air traffic is quieter, the sky is breathing.

The field is whispering in new ease and in new challenges
…this Quieting for some is emotionally challenging and for

others an awe-
some spiritual miracle.
With laughter and humming we are giving language and

life to our
individual and collective transition.

Our transition into deeper release,
deeper nowness,
deeper selfness
and deeper life witnessing.
Hopefully with
a deeper gratitude for our own inner quiet being.

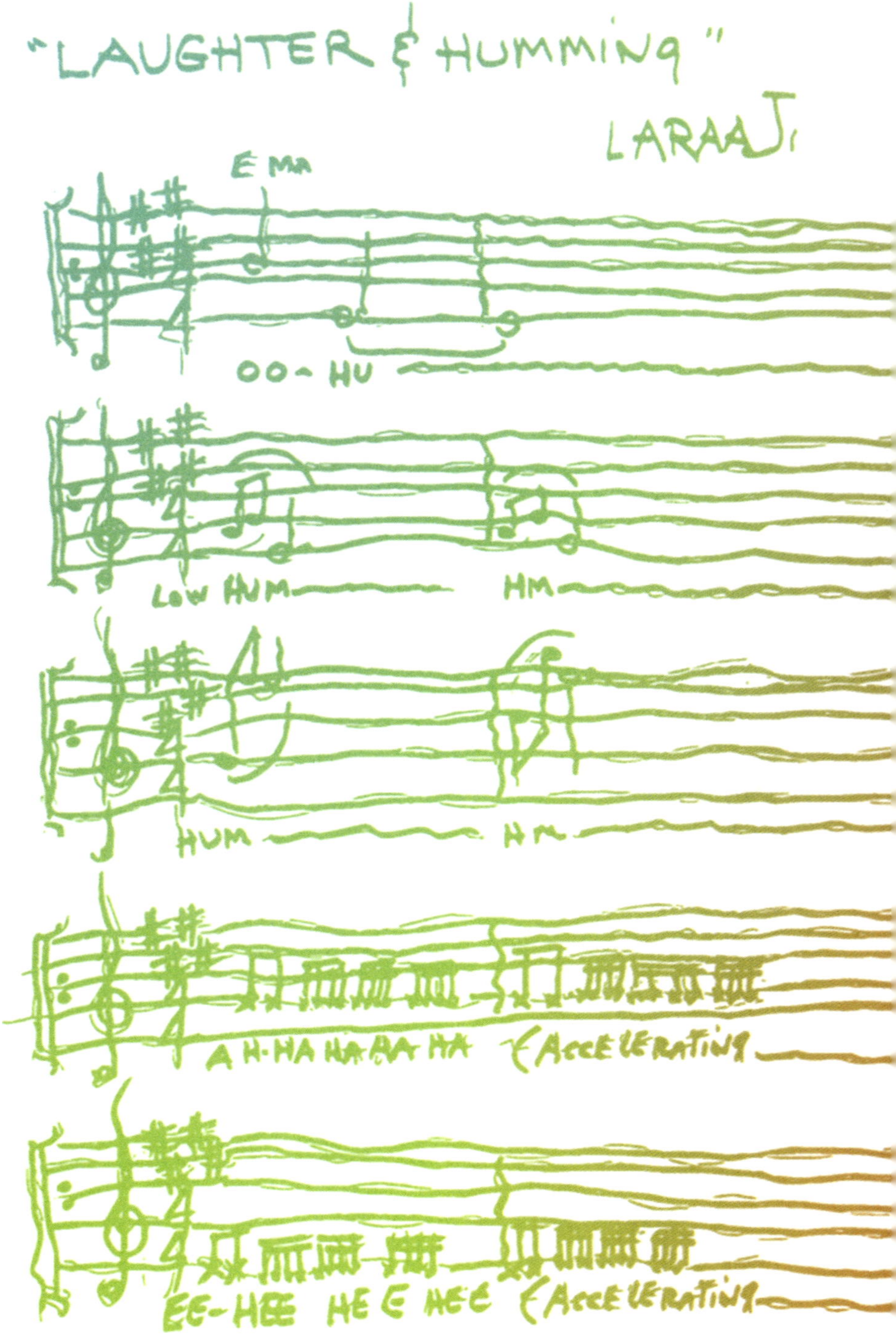

"LAUGHTER & HUMMING"
LARAAJI
E MA
OO~HU
LOW HUM
HM
HUM
HM
AH-HA HA HA HA (ACCELERATING
EE-HEE HE E HEE (ACCELERATING

BMA
OO ~ HU
COO
CROON
LAUGH
HIM
H M
HUM
H M
H M
HUM
H M
LAUGHTER/CHUCKLE)
LAUGH
CHUCKLE
GIGGLE (HIGH VOCAL)
GIGGLE

Postlude

Thomas Lecocq[1], Stephen P. Hicks[2], Koen Van Noten[1], Kasper van Wijk[3], Paula Koelemeijer[4]
Theresia Apoloner[9], Mario Arroyo-Solórzano[10], Jelle D. Assink[11], Pinar Büyükakpınar[12,13]
Chaves[18], David G. Cornwell[19], David Craig[20], Olivier F. C. den Ouden[11,21], Jordi Diaz[22], Stefani[e]
Dimitrios Giannopoulos[27,28], Steven J. Gibbons[29], Társilo Girona[30], Bogdan Grecu[31], Mar[c]
Mohammadreza Jamalreyhani[37,13], Alan Kafka[38], Mathijs R. Koymans[11,21], Celeste R. Labedz[39]
William Minarik[45,46], Louis Moresi[44], Víctor H. Márquez-Ramírez[5], Martin Möllhoff[20], Ian M[.]
Retailleau[54,55], Annukka E. Rintamäki[7], Claudio Satriano[54], Martha K. Savage[56], Shahar Shani[-]
Shiba Subedi[33], Mathilde B. Sørensen[59], Taka'aki Taira[60], Mar Tapia[61], Fatih Turhan[12], Ben va[n]
Joachim Wassermann[43], Han Xiao[66]

Global quieting of high-frequency seismic noise due to COVID-19 pandemic lockdown measures

Article and figures originally published in *Science*
11 Sep 2020: Vol. 369, Issue 6509, pp. 1338-1343
Reprinted with permission from AAAS.

[1]Seismology-Gravimetry, Royal Observatory of Belgium, Brussels, Belgium. [2]Department of Earth Science and Engineering, Imperial College London[,]
Egham, UK. [5]Centro de Geociencias, Universidad Nacional Autónoma de México, Campus Juriquilla, Querétaro, Mexico. [6]Swiss Seismological Service[,]
Geological Survey, Albuquerque, NM, USA. [9]Zentralanstalt für Meteorologie und Geodynamik (ZAMG), Vienna, Austria. [10]Escuela Centroamerican[a]
De Bilt, Netherlands. [12]Kandilli Observatory and Earthquake Research Institute, Boğaziçi University, Istanbul, Turkey. [13]GFZ German Research[,]
Catania, Italy. [15]Istituto Nazionale di Geofisica e Vulcanologia, Osservatorio Etneo, Catania, Italy. [16]Bensberg Observatory, University of Cologne[,]
Seismological Observatory of Costa Rica at Universidad Nacional (OVSICORI-UNA), Heredia, Costa Rica. [19]Department of Geology and Geophysics[,]
Ireland. [21]Department of Geoscience and Engineering, Delft University of Technology, Delft, Netherlands. [22]Geosciences Barcelona, CSIC, Barcelona[,]
Athens, Athens, Greece. [25]Noise Department, Brussels Environment, Brussels-Capital Region, Belgium. [26]Observatorio San Calixto, La Paz, Bolivia[.]
Hellenic Mediterranean University, Chania, Greece. [29]Norges Geotekniske Institutt, Oslo, Norway. [30]Geophysical Institute, University of Alask[a,]
Université de Strasbourg, CNRS, EOST UMS830, Strasbourg, France. [33]Institute of Earth Sciences, Faculty of Geosciences and Environment[,]
Peru, Lima, Peru. [36]Department of Geosciences, Princeton University, Princeton, NJ, USA. [37]Institute of Geophysics, University of Tehran, Tehran[,]
California Institute of Technology, Pasadena, CA, USA. [40]Geophysics Department, Stanford University, Stanford, CA, USA. [41]SETI Institute, Mountain[n]
Canada. [43]Ludwig-Maximilians-Universität München, Munich, Germany. [44]Research School of Earth Sciences, Australian National University[,]
Montréal, QC, Canada. [47]Raspberry Shake, S.A., Boquete, Chiriqui, Panama. [48]Department of Earth and Climate Science, University of Maine[,]
Center for Geodynamics and Seismology, Walferdange, Grand Duchy of Luxembourg. [52]National Centre for Earth Observation, Department of[,]
de Paris, Paris, France. [55]Observatoire Volcanologique du Piton de la Fournaise, Institut de Physique du Globe de Paris, La Plaine des Cafres[,]
Geology, University of Patras, Patras, Greece. [58]Department of Electrical and Electronic Engineering, Imperial College London, South Kensington[,]
California Berkeley, Berkeley, CA, USA. [61]Laboratori d'Estudis Geofísics Eduard Fontserè, Institut d'Estudis Catalans (LEGEF-IEC), Barcelona, Spain[.]
Cornwall, UK. [64]Institut de Physique du Globe de Strasbourg, UMR 7516, Université de Strasbourg/EOST, CNRS, Strasbourg, France. [65]Department[,]
CA, USA. *Corresponding author. Email: thomas.lecocq@seismology.be

aphael S. M. De Plaen[5], Frédérick Massin[6], Gregor Hillers[7], Robert E. Anthony[8], Maria-
ndrea Cannata[14,15], Flavio Cannavo[15], Sebastian Carrasco[16], Corentin Caudron[17], Esteban J.
onner[23], Christos P. Evangelidis[24], Läslo Evers[11,21], Benoit Fauville[25], Gonzalo A. Fernandez[26],
runberg[32], György Hetényi[33], Anna Horleston[34], Adolfo Inza[35], Jessica C. E. Irving[34,36],
ric Larose[17], Nathaniel J. Lindsey[40], Mika McKinnon[41,42], Tobias Megies[43], Meghan S. Miller[44],
esbitt[47,48], Shankho Niyogi[49], Javier Ojeda[50], Adrien Oth[51], Simon Proud[52], Jay Pulli[53,38], Lise
admiel[21], Reinoud Sleeman[11], Efthimios Sokos[57], Klaus Stammler[23], Alexander E. Stott[58],
er Pluijm[62], Mark Vanstone[63], Jerome Vergne[64], Tommi A. T. Vuorinen[7], Tristram Warren[65],

Human activity causes vibrations that propagate into the ground as high-frequency seismic waves. Measures to mitigate the coronavirus disease 2019 (COVID-19) pandemic caused widespread changes in human activity, leading to a months-long reduction in seismic noise of up to 50%. The 2020 seismic noise quiet period is the longest and most prominent global anthropogenic seismic noise reduction on record. Although the reduction is strongest at surface seismometers in populated areas, this seismic quiescence extends for many kilometers radially and hundreds of meters in depth. This quiet period provides an opportunity to detect subtle signals from subsurface seismic sources that would have been concealed in noisier times and to benchmark sources of anthropogenic noise. A strong correlation between seismic noise and independent measurements of human mobility suggests that seismology provides an absolute, real-time estimate of human activities.

ondon, UK. [3]Department of Physics, University of Auckland, New Zealand. [4]Department of Earth Sciences, Royal Holloway University of London,
TH Zurich, Zurich, Switzerland. [7]Institute of Seismology, University of Helsinki, Helsinki, Finland. [8]Albuquerque Seismological Laboratory, U.S.
e Geología, Universidad de Costa Rica, San José, Costa Rica. [11]R&D Seismology and Acoustics, Royal Netherlands Meteorological Institute (KNMI),
entre for Geosciences, Potsdam, Germany. [14]Dipartimento di Scienze Biologiche, Geologiche e Ambientali, Università Degli Studi di Catania,
ologne, Germany. [17]Université. Grenoble Alpes, Université Savoie Mont Blanc, CNRS, IRD, IFSTTAR, ISTerre, Grenoble, France. [18]Volcanological and
chool of Geosciences, University of Aberdeen, King's College, Aberdeen, UK. [20]Dublin Institute for Advanced Studies, Geophysics Section, Dublin,
pain. [23]Federal Institute for Geosciences and Natural Resources (BGR), Hannover, Germany. [24]Institute of Geodynamics, National Observatory of
Seismotech S.A., Athens, Greece. [28]Laboratory of Geophysics & Seismology, Department of Environmental & Natural Resources Engineering,
airbanks, Fairbanks, AK, USA. [31]National Institute for Earth Physics, Magurele, Romania. [32]Réseau National de Surveillance Sismique (RENASS),
niversity of Lausanne, Lausanne, Switzerland. [34]School of Earth Sciences, University of Bristol, Queen's Road, Bristol, UK. [35]Instituto Geofisico del
an. [38]Weston Observatory, Department of Earth and Environmental Sciences, Boston College, Weston, MA, USA. [39]Seismological Laboratory,
iew, CA, USA. [42]Faculty of Science, Department of Earth, Ocean and Atmospheric Sciences, University of British Columbia, Vancouver, BC,
anberra, ACT, Australia. [45]Department of Earth and Planetary Sciences, McGill University, Montréal, QC, Canada. [46]GEOTOP Research Centre,
rono, ME, USA. [49]University of California, Riverside, CA, USA. [50]Departamento de Geofísica, Universidad de Chile, Santiago, Chile. [51]European
hysics, University of Oxford, Oxford, UK. [53]Raytheon BBN Technologies, Arlington, VA, USA. [54]Université de Paris, Institut de Physique du Globe
rance. [56]School of Geography, Environment and Earth Sciences, Victoria University of Wellington, Wellington, New Zealand. [57]Department of
ampus, London, UK. [59]Department of Earth Science, University of Bergen, Bergen, Norway. [60]Berkeley Seismological Laboratory, University of
Department of Earth and Environmental Sciences, University of Michigan, Ann Arbor, MI, USA. [63]Geology Department, Truro School, Truro,
f Physics, University of Oxford, Oxford, UK. [66]Department of Earth Science and Earth Research Institute, University of California, Santa Barbara,

1. J. N. Brune, J. Oliver, *Bull. Seismol. Soc. Am.* 49, 349-353 (1959).

2. R. K. Cessaro, *Bull. Seismol. Soc. Am.* 84, 142-148 (1994).

3. N. M. Shapiro, M. Campillo, *Geophys. Res. Lett.* 31, L07614 (2004).

4. D. E. McNamara, R. P. Buland, *Bull. Seismol. Soc. Am.* 94, 1517-1527 (2004).

5. J. C. Groos, J. R. Ritter, *Geophys. J. Int.* 179, 1213-1231 (2009).

6. C. M. Boese, L. Wothespoon, M. Alvarez, P. Malin, *Bull. Seismol. Soc. Am.* 105, 285-299 (2015).

Seismometers record signals from more than just earthquakes: Interactions between the solid Earth and fluid bodies, such as ocean swell and atmospheric pressure,[1,2] are now commonly used to image and monitor the subsurface.[3] Human activity is a third source of seismic signal. Nuclear explosions and fluid injection or extraction result in impulsive signals, but everyday human activity is recorded as a near-continuous signal, especially on seismometers in urban environments. These complicated signals are the superposition of a wide variety of activities happening at different times and places at or near Earth's surface but are typically stronger during the day than at night, weaker on weekends than weekdays, and stronger near population centers than sparsely inhabited areas.[4-7] Seismometers in urban environments are important to maximize the spatial coverage of seismic networks and to warn of local geologic hazards,[8] even though anthropogenic seismic noise degrades their capability to detect transient signals associated with earthquakes and volcanic eruptions. Therefore, it is vital to understand urban seismic sources, but studies have been limited to confined areas or distinct events, such as road traffic,[9,10] public transport,[7,11] and "football quakes."[11,12] Broad analysis of the long-term global anthropogenic seismic wavefield has been lacking. The impact of large, coherent changes in human behavior on seismic noise is unknown, as is how far it propagates

7. D. N. Green, I. D. Bastow, B. Dashwood, S. E. Nippress, *Seismol. Res. Lett.* 88, 113-124 (2017).

8. C. L. Ashenden et al., *Nat. Hazards* 59, 507-528 (2011).

9. N. Riahi, P. Gerstoft, *Geophys. Res. Lett.* 42, 2674-2681 (2015).

10. N. J. Lindsey et al., *Geophys. Res. Lett.* 47, e2020GL089931 (2020).

11. J. Díaz, M. Ruiz, P. S. Sánchez-Pastor, P. Romero, *Sci. Rep.* 7, 15296 (2017).

12. P. Denton, S. Fishwick, V. Lane, D. Daly, *Seismol. Res. Lett.* 89, 1902-1907 (2018).

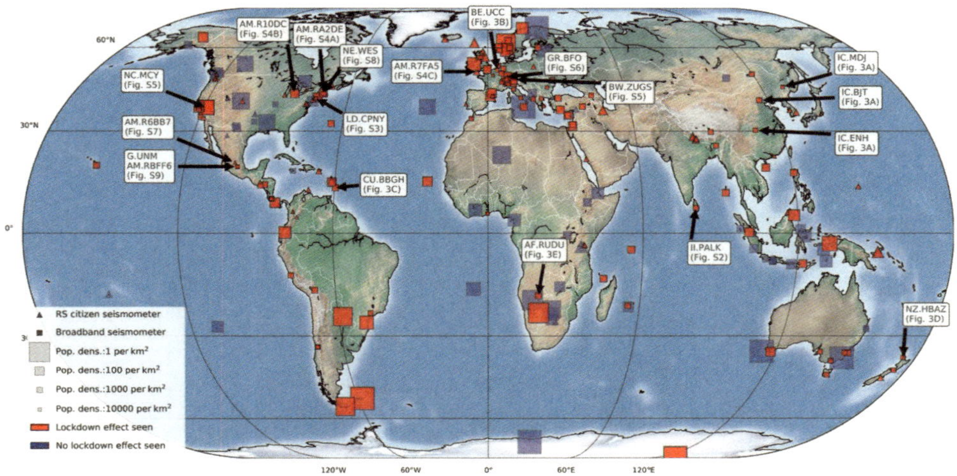

Fig. 1. Locations of analyzed seismic stations throughout the world. The map shows locations of the 268 global seismic stations with usable data (e.g., no long data gaps, working sensors) that we analyzed. Lockdown effects were observed (red) at 185 of 268 stations. Symbol size is scaled by the inverse of population density (30) to emphasize stations located in remote areas. The labeled stations are discussed in detail in the text.

and whether seismic recordings offer a coarse proxy for monitoring human activity patterns. Answering these questions has proven challenging because datasets are large, monitoring networks are heterogeneous, and the many possible noise sources likely vary spatially and overlap in time.[13]

The coronavirus disease 2019 (COVID-19) outbreak was declared a global health emergency in January 2020[14] and a pandemic in March 2020 by the World Health Organization. The outbreak resulted in emergency measures to reduce the basic reproduction rate of the virus,[15] beginning in China and Italy and then followed by most countries. These measures disrupted

13. D. Wilson *et al.*, *Bull. Seismol. Soc. Am.* 92, 3335-3342 (2002).

14. C. Sohrabi et al., *Int. J. Surg.* 76, 71-76 (2020).

15. R. M. Anderson, H. Heesterbeek, D. Klinkenberg, T. D. Hollingsworth, *Lancet* 395, 931-934 (2020).

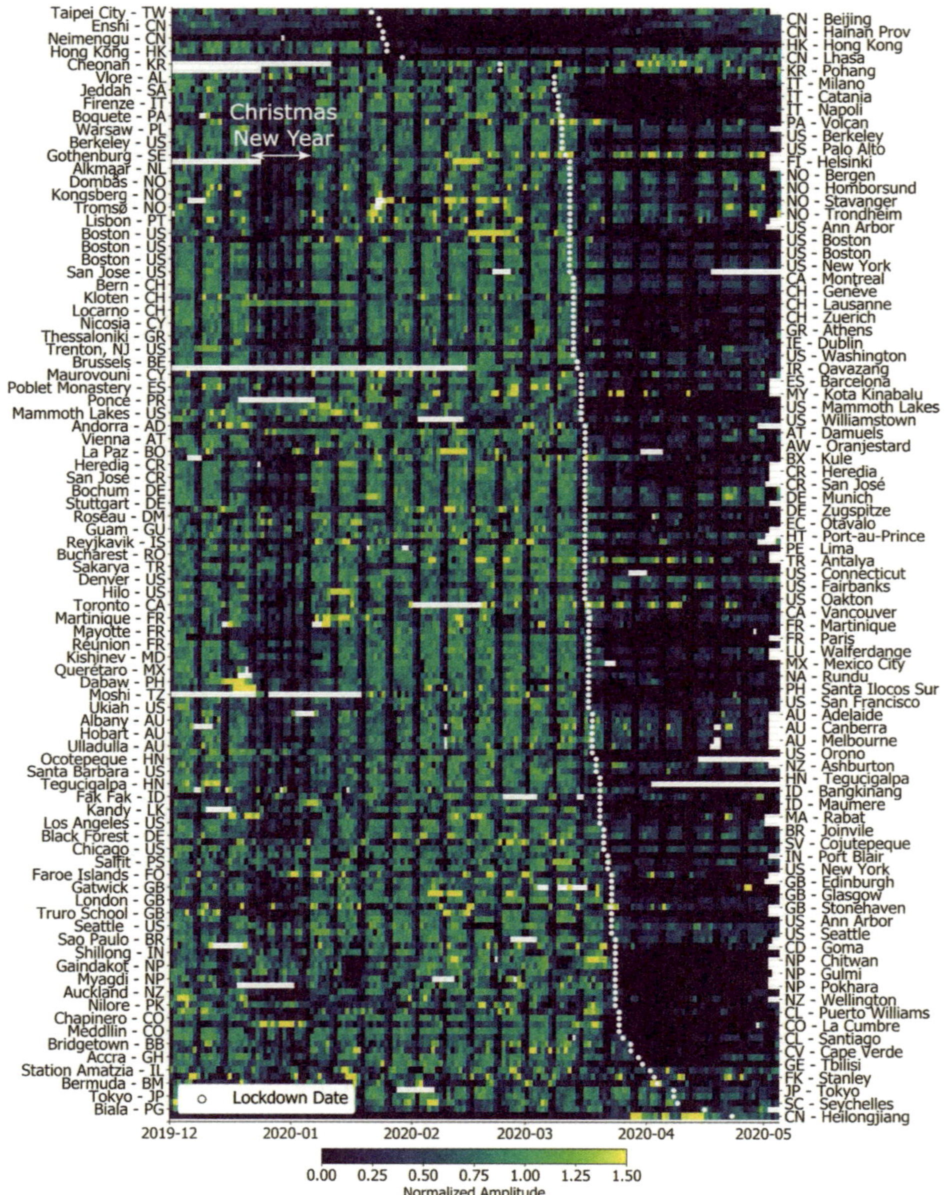

Fig. 2. Global temporal changes in seismic noise. Global daily median hiFSAN is depicted (24), normalized to percentage variation of the baseline before lockdown measures and sorted by lockdown date. Each line of the image corresponds to one seismic station. Data gaps are shown in white. Location and country code are indicated for each station; see fig. S1 for network and station codes.

16. M. Nicola *et al.*, *Int. J. Surg.* 77, 206-216 (2020).

17. T. Laing, *Extr. Ind. Soc.* 7, 580-582 (2020).

18. A. Hoque, F. A. Shikha, M. W. Hasanat, I. Arif, A. B. A. Hamid, *Asian J. Multidiscip. Stud.* 3, 52-58 (2020).

19. M. U. G. Kraemer *et al.*, *Science* 368, 493-497 (2020).

20. H. Tian *et al.*, *Science* 368, 638-642 (2020).

21. J. Zhang *et al.*, *Science* 368, 1481-1486 (2020).

22. M. Chinazzi *et al.*, *Science* 368, 395-400 (2020).

23. M. Bauwens *et al.*, *Geophys. Res. Lett.* 47, 11 (2020).

24. Materials and methods and network citations are available as supplementary materials.

25. R. E. Anthony, A. T. Ringler, D. C. Wilson, E. Wolin,. *Seismol. Res. Lett.* 90, 219-228 (2019).

26. H. Xiao, Z. C. Eilon, C. Ji, T. Tanimoto, *Sesimol. Res. Lett.* 10.1785/0220200147 (2020).

27. P. Poli, J. Boaga, I. Molinari, V. Cascone, L. Boschi, *Sci. Rep.* 10, 9404 (2020).

28. H. Lau *et al.*, *J. Travel Med.* 27, taaa037 (2020).

social and economic behavior,[16] industrial production,[17] and tourism.[18] In this paper, we use the term "lockdown" to broadly encompass many types of emergency measures, such as full quarantine [e.g., in Wuhan, China[19–21]], enforced physical distancing (e.g., in Italy and the United Kingdom), travel restrictions,[22] widespread closure of services and industry, and any other emergency measures. These major changes to daily life provide an opportunity to study their environmental impacts, such as reductions in nitrous oxide emissions in the atmosphere.[23] Recordings of human-generated seismic vibrations that travel through the solid Earth provide insights into the dynamics of pandemic lockdowns.

We assessed the effects of COVID-19 lockdowns on high-frequency (4 to 14 Hz) seismic ambient noise (hiFSAN).[24] We compiled a global seismic noise dataset using vertical-component seismic waveform data from 337 broadband and individually operated citizen seismometer stations,[24] such as Raspberry Shake instruments (RSs), with a self-noise well below the ground motion generated by anthropogenic noise[25] and flat responses in the target frequency band (Fig. 1). We obtained usable data (e.g., no large data gaps, working sensors) from 268 stations and detected pronounced reductions in hiFSAN during local lockdown measures at 185 stations (Fig. 2). Periods that are often seismically quiet include weekends, as well as the Christmas and New Year holidays for locations where they are celebrated. Notably, we found a near-global reduction in noise, commencing in China in late January 2020,[26] followed by Italy,[26, 27] the whole of Europe, and the rest of the world in March to April 2020. This period of reduced noise lasted longer and was often quieter than the Christmas–to–New Year period.

In China (Fig. 3A), the COVID-19 outbreak and subsequent emergency measures occurred during the Chinese New Year (CNY). In Enshi, a city located in Hubei province where the outbreak began,[28] hiFSAN in 2020 clearly diverged from the normal annual reduction during CNY. The hiFSAN level remained at a minimum, demarcated by the start and end of quarantine in Hubei, for several weeks after CNY. Although the quarantine measures in Beijing were less strict, local hiFSAN reductions were more pronounced and lasted longer than in recent years. As of the end date of our analysis, Beijing has still not reached the average hiFSAN level of previous years, which suggests that the impact of COVID-19 is continuing to restrict anthropogenic noise there. We noticed a later hiFSAN lockdown reduction in April 2020 in Heilongjiang (Fig. 3A), in northeast China, near the Russian border.

24. Materials and methods and network citations are available as supplementary materials.

Although we observed seismic effects of lockdown in areas with low population density estimates (<1 person per km^2; Fig. 1), the strongest hiFSAN reduction occurred in populated environments. For a permanent seismic station in Sri Lanka, a 50% reduction in hiFSAN occurred after lockdown, which is the strongest we observed in the available data from that station since at least July 2013 (fig. S2). In Central Park, New York, on Sunday nights, hiFSAN was 10% lower during the lockdown than before this period (fig. S3).

Seismic networks in populated areas enable us to correlate hiFSAN with other human activity measurements, such as audible recordings and flight data.[24] At a surface station in Brussels, Belgium (Fig. 3B), we found a 33% reduction in hiFSAN after lockdown. We compared this noise level with data from a nearby microphone, located close to a major road, that mainly records audible traffic noise. We found a high correlation between pre-lockdown hiFSAN and audible noise, both showing characteristic diurnal and weekly changes. However, during lockdown, audible noise reductions were more pronounced, which suggests that seismometers are sensitive to a wide distribution of seismic sources, not just nearby traffic. Audible and hiFSAN levels then gradually increased after April 2020. Independent mobility data[24] provide insights into what caused these changes. Mobility correlates with hiFSAN at lockdown, with correlation coefficients >0.8,[24] except for time spent

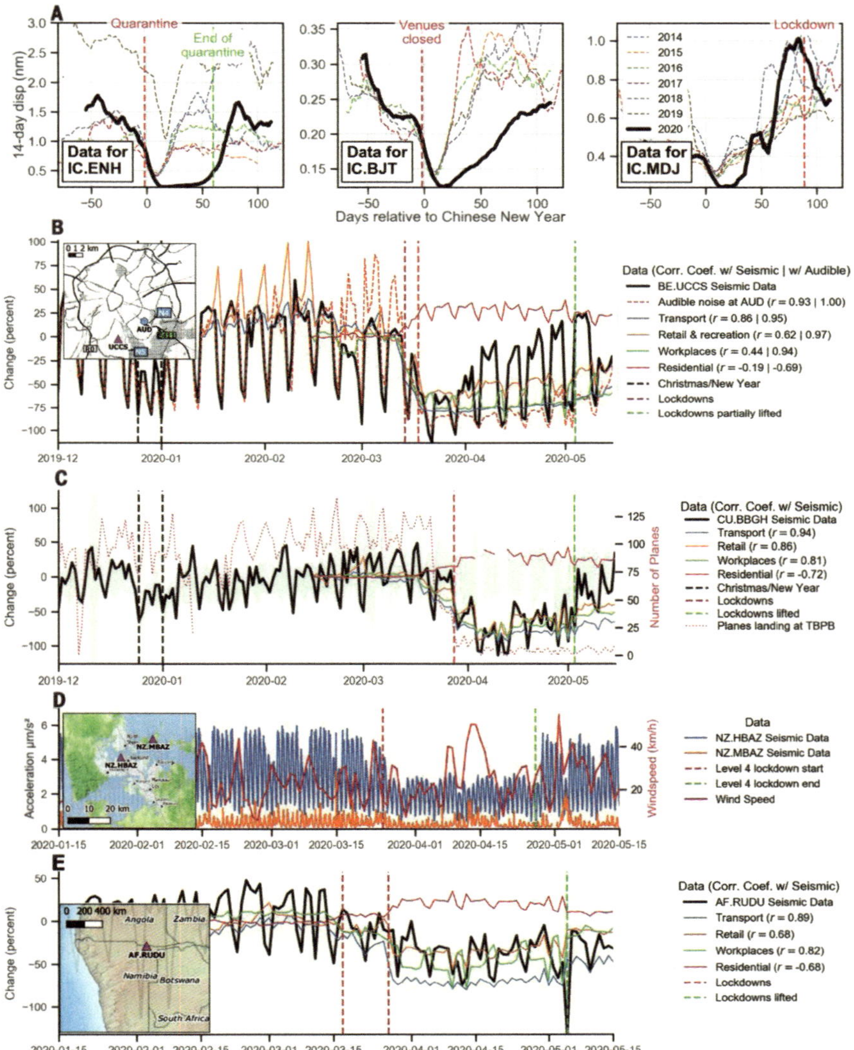

Fig. 3. Regional examples of the 2020 seismic noise quiet period. The examples show different features of the lockdown seismic signal changes in regional settings. We filtered the hiFSAN data between 4 and 14 Hz and present temporal changes as displacement (A), acceleration (D), or percentage change relative to the baseline before lockdown [(B), (C), and (E)], with the panels in (A) also relative to the baseline of corresponding time periods in previous years. Individual seismic stations are identified by codes in "network. station" format (IC. ENH, BE. UCCS, etc.). The keys in (B) to (E) include correlation coefficients (r) with mobility data (24). (A) Lockdown effects at three stations in China compared with the Chinese New Year holiday in previous years. (B) Lock- down effects on hiFSAN compared with audible environmental noise and independent mobility data in Brussels, Belgium. (C) Lockdown effect in Barbados compared with noise levels of the past decade (gray shading) and correlation with local flight data at the Grantley Adams International Airport (TBPB) (24). (D) Lockdown noise reduction recorded on borehole seismometers in Auckland, New Zealand. (E) Lockdown noise reduction in a region of low population density in Rundu, Namibia.

at places of residence (Google's "residential" category), which is expected given the increased number of people spending more time at home because of government restrictions.

Citizen seismometers provide a different urban ground motion dataset, with denser coverage in some places. Large hiFSAN drops occurred particularly at schools and universities after lockdown-related closures [e.g., in Boston and Michigan (United States) and Cornwall (United Kingdom); fig. S4]. The hiFSAN level was even 20% lower than during school holidays, which indicates sensitivity to the environment outside of the school.

The pandemic has also affected tourism — for example, during the holiday season in the Caribbean. In Barbados (Fig. 3C), hiFSAN decreased by ~45% after lockdown on 28 March 2020 through April 2020 and stayed ~50% below levels observed in previous years for the same period. However, seismic noise levels began to decrease 1 to 2 weeks before a local curfew was implemented. Local flight data[24] indicate that travel to Barbados started decreasing after 21 March 2020, and the overall reduction in hiFSAN might have been partly due to tourists repatriating. We also observed noise reductions due to decreased tourist activity at ski resorts in Europe (Zugspitze in Germany) and the United States (Mammoth Mountain in California) (fig. S5).

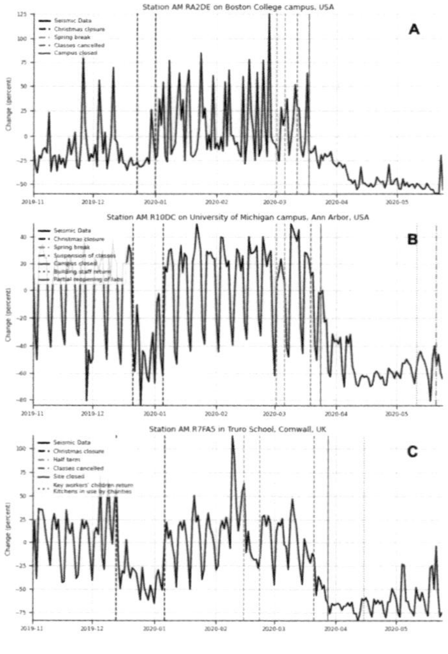

Fig. S4. Temporal changes in hiFSAN at schools and universities. (A) Boston College (US). (B) University of Michigan campus (US). (C) Truro School (Cornwall, UK). Time periods like the Christmas break and Spring break / half term are easily identified as periods of lower seismic noise. Note the large drop in hiFSAN up to more than 50 % change after classes are cancelled and campuses are closed.

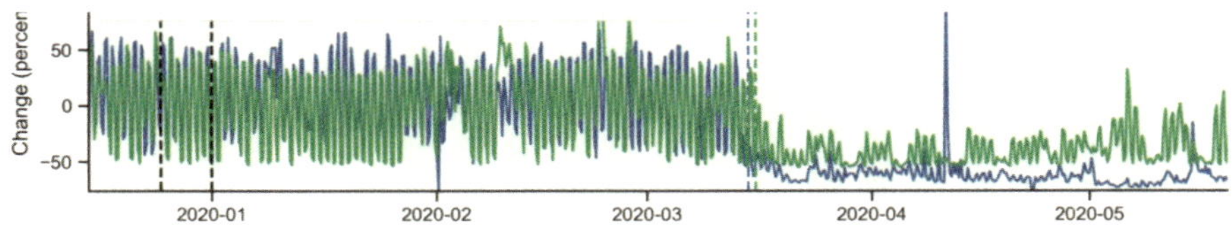

Fig. S5. Lockdown effects on seismic noise in touristic ski resort areas.
Temporal changes of hiFSAN at two touristic ski resort areas: Mammoth Mountain (USA - station NC.MCY) and Zugspitze (Germany, station BW.ZUGS), showing the effect of the complete stop of the resorts' lifts or cog-wheel railway

6. C. M. Boese, L. Wothespoon, M. Alvarez, P. Malin, *Bull. Seismol. Soc. Am.* 105, 285-299 (2015).

8. C. L. Ashenden et al., *Nat. Hazards* 59, 507-528 (2011).

29. S. Sherburn, B. J. Scott, J. Olsen, C. Miller, *N. Z. J. Geol. Geophys.* 50, 1-11 (2007).

Although we observed lockdown effects most prominently at surface stations, we also detected them underground. In New Zealand, seismometers installed in boreholes (to minimize the effects of anthropogenic noise) monitor potential hazards associated with the Auckland Volcanic Field.[6,8,29] Station HBAZ is 380 m below the city, whereas MBAZ is at 98 m of depth, 14 km from the city center on the uninhabited Motutapu Island (Fig. 3D). The hiFSAN level at both stations varied between weekdays and weekends before the lockdown, which suggests that both are sensitive to anthropogenic activity. Although the island station is quieter overall, the lockdown instigated a reduction in hiFSAN by a factor of 2 for both stations. We attribute the remaining hiFSAN maxima on the island (mid-April 2020 and early May 2020) to strong winds and high waves. On 27 April 2020, New Zealand lifted restrictions, with hiFSAN increasing to the pre-lockdown levels.

The reduction of hiFSAN was weaker in less populated areas such as Rundu, located along the Namibia-Angola border (Fig. 3E). After COVID-19 was confirmed in Namibia, an emergency was declared on 17 March 2020 to restrict mobility, followed by full lockdown on 27 March 2020. These measures are reflected in the >25% hiFSAN reduction compared with pre-lockdown levels. Despite Rundu having a population roughly one-eighth and one-fifth as dense as those of Brussels and Auckland, respectively,[30] we observed a similarly high correlation between seismic and mobility data. The Black Forest Observatory in Germany is an even more remote station, located 150 to 170m below the surface in crystalline bedrock. Although this station is considered a reference laboratory with low noise overall,[31] we detected a small hiFSAN reduction during lockdown nights (fig. S6), corresponding to the lowest hiFSAN since at least 25 December 2015.

Here we have provided a global-scale analysis of high-frequency anthropogenic seismic noise. Global median hiFSAN dropped by as much as 50% during March to May 2020 (Fig. 4). The length and quiescence of this period represent the longest and most coherent global seismic noise reduction in recorded history, emphasizing how human activities affect the solid Earth. A globally high correlation exists between changes in hiFSAN and population mobility,[24] with correlations exceeding 0.9 for many categories.

30. Center for International Earth Science Information Network CIESIN Columbia University, Gridded population of the world, version 4 (GPWv4): Population density. Revision 11, accessed 2 June 2020 (2018); https://doi.org/10.7927/H49C6VHW.

31. W. Zürn et al., Geophys. J. Int. 171, 780-796 (2007).

24. Materials and methods and network citations are available as supplementary materials.

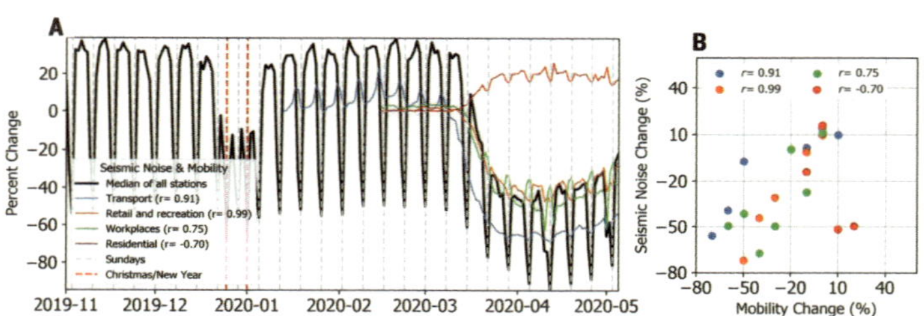

Fig. 4. Global changes in seismic noise compared with population mobility trends.
(A) Comparison between temporal changes in global daily median hiFSAN based on data from the 185 stations that observed lockdown effects and population mobility changes (24). (B) Scatter plot to illustrate the correlation between the binned (10% bins) time series of seismic noise changes and all categories of mobility data in (A). Percentage changes are expressed relative to a pre-lockdown baseline. All categories show a strong positive correlation, apart from time spent in residential premises, which is anticorrelated.

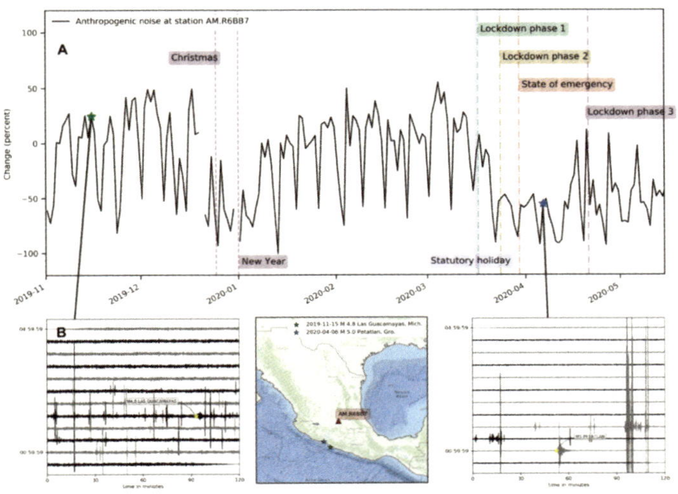

Fig. S7. Increased earthquake detection capability in Mexico.
(A) Temporal changes in hiFSAN at Raspberry Shake station AM.R6BB7 (city center of Querétaro, Mexico). (B) Earthquakes at Las Guacamayas (15/11/2019, M4.8 and 28 km depth) and Petatlan (07/04/2020, M5.0 and 15 km depth) recorded before (left column) and after the lockdown (right). The insert in the center indicates the locations of the earthquakes and Raspberry Shake station.

4. D. E. McNamara, R. P. Buland, *Bull. Seismol. Soc. Am.* 94, 1517-1527 (2004).

32. B. Gutenberg, C. F. Richter, *Bull. Seismol. Soc. Am.* 34, 185-188 (1944).

33. Z. E. Ross, D. T. Trugman, E. Hauksson, P. M. Shearer, *Science* 364, 767-771 (2019).

This distinct low-noise period will help optimize seismic monitoring efforts.[4] The ability to analyze the full spectrum of seismogenic behavior, including the smallest earthquakes, is essential for monitoring fault dynamics over seismic cycles, as well as for earthquake forecasting and seismic hazard assessment. Small earthquakes should dominate datasets,[32] but typical operational catalogs using amplitude-based detection do not include many of the smallest earthquakes.[33] This detection issue is especially problematic in populated areas, where anthropogenic noise energy interferes with earthquake signals. This problem is exemplified by recordings of a moment magnitude 5.0 earthquake at 15 km of depth southwest of Petatlan, Mexico, during lockdown (fig. S7). An earthquake with this magnitude and source mechanism that occurs during the daytime would typically be observed at stations in urban environments only if the signal was filtered. However, the reduction of seismic noise by ~40% during lockdown made this event visible, without any filtering required, at a RS station in Querétaro city, 380 km away. Low noise levels during COVID-19 lockdowns could thus allow detection of signals from previously unrecognized sources in areas with incomplete seismic catalogs. Such newly identified signals could be used as distinct templates[32] for finding similar waveforms in noisier data before and after lockdown. This approach also works for tremor signals that are masked by anthropogenic noise yet vital for monitoring potential volcanic unrest.[6] Although broadband sensors in rural environments are less affected by anthropogenic noise, any densification of and reliance on low-cost sensors in urban areas, such as RSs and low-cost accelerometers,[34] will require a better understanding of anthropogenic noise sources to suppress false detections. As populations increase globally, more people become exposed to potential natural and induced geohazards.[35] Urbanization will increase anthropogenic noise in exposed areas, further complicating seismic monitoring efforts. The ability to characterize and minimize anthropogenic noise is becoming increasingly important for accurate detection and imaging of seismic signatures of potentially harmful subsurface hazards.

Anthropogenic seismic noise is thought to be dominated by noise sources <1 km away from detectors.[5–7, 11, 36] Because population mobility generates time-varying loads that radiate energy through the shallow subsurface as Rayleigh waves,[11] local effects such as construction sites and heavy machinery can affect individual stations. However, the 2020

32. B. Gutenberg, C. F. Richter, *Bull. Seismol. Soc. Am.* 34, 185-188 (1944).

6. C. M. Boese, L. Wothespoon, M. Alvarez, P. Malin, *Bull. Seismol. Soc. Am.* 105, 285-299 (2015).

34. E. S. Cochran, *Nat. Commun.* 9, 2508 (2018).

35. G. J. H. McCall, *Geol. Soc. London Eng. Geol. Spec. Publ.* 15, 309-318 (1998).

5. J. C. Groos, J. R. Ritter, *Geophys. J. Int.* 179, 1213-1231 (2009).

6. C. M. Boese, L. Wothespoon, M. Alvarez, P. Malin, *Bull. Seismol. Soc. Am.* 105, 285-299 (2015).

7. D. N. Green, I. D. Bastow, B. Dashwood, S. E. Nippress, *Seismol. Res. Lett.* 88, 113-124 (2017).

11. J. Díaz, M. Ruiz, P. S. Sánchez-Pastor, P. Romero, *Sci. Rep.* 7, 15296 (2017).

36. M. Lehujeur, J. Vergne, J. Schmittbuhl, A. Maggi, *Geotherm. Energy* 3, 3 (2015).

seismic noise quiet period reveals that when considering multiple stations or whole networks over longer time scales, the anthropogenic seismic wavefield affects large areas. With denser networks and more citizen sensors in urban environments, additional features of the seismic noise, rather than just amplitude, will become usable and will help identify different anthropogenic noise sources.[10,37] Characterizing these sources will be useful for imaging the shallow subsurface in three dimensions in urban areas by using high-frequency anthropogenic ambient noise.[38,39] Our finding of a distributed noise field is supported by strong correlations with independent mobility data (Fig. 4). In contrast to mobility data, publicly available data from existing seismometer networks provide an objective absolute baseline of human activity levels. Therefore, hiFSAN can serve as a near–real-time technique for monitoring anthropogenic activity patterns, with fewer potential privacy concerns than those raised by mobility data collection. In addition, although industrial activities may not be captured in mobility data, they may produce a seismic noise signature. The 2020 seismic quiet period is a baseline for using seismic properties[36] to identify and isolate the sources contributing to the anthropogenic noise wavefield, especially when combined with data indicative of human behavior. Seismic observations of human activity during COVID-19 lockdowns have enabled us to assess the impact of mitigation policies — particularly the time to establish and recover from lockdowns — on daily life. As such, hiFSAN may provide important constraints for future health and behavioral science studies.

10. N. J. Lindsey et al., *Geophys. Res. Lett.* 47, e2020GL089931 (2020).

37. G. Hillers, M. Campillo, Y.-Y. Lin, K.-F. Ma, P. Roux, *J. Geophys. Res.* 117, B06301 (2012).
38. M. Picozzi, S. Parolai, D. Bindi, A. Strollo, *Geophys. J. Int.* 176, 164-174 (2009).

39. F. Brenguier *et al.*, *Geophys. Res. Lett.* 46, 9529-9536 (2019).

36. M. Lehujeur, J. Vergne, J. Schmittbuhl, A. Maggi, *Geotherm. Energy* 3, 3 (2015).

40. T. Lecocq *et al.*, ThomasLecocq/2020_Science_GlobalQuieting: First Release - v1.0, Zenodo (2020); https://doi.org/10.5281/zenodo.3944739.

Reprinted with Permission from *Science* 11 Sep 2020: Vol. 369, Issue 6509, pp. 1338-1343 DOI: 10.1126/science.abd2438

SUPPLEMENTARY MATERIALS science.sciencemag.org/content/369/6509/1338/suppl/DC1 Materials and Methods Supplementary Text Figs. S1 to S9 Tables S1 to S3 References (41-56) Movie S1

10 June 2020; accepted 14 July 2020 Published online 23 July 2020 10.1126/science.abd2438

Every Wiggle Counts

Stuart Hyatt   Let's start with the basics. What is a seismometer and how does it work exactly?

Debi Kilb, PhD
Project Scientist, Institute of Geophysics and Planetary Physics
Scripps Institution of Oceanography
University of California San Diego

Debi Kilb   A seismometer is an instrument that measures ground motion. A typical seismometer is composed of three components that measure north-south motion, east-west motion, and vertical up-down motion. The terms seismometer and seismograph are often confused: a seismometer is the recording instrument and a seismograph is the data wiggle recorded by the seismometer.

The inner workings of the seismometer have evolved over time. Some of the more beautiful seismometers, what many true aficionados call art, were elaborate sculptures that, when shaken, dropped large marble balls indicating significant ground motions had occurred.

Later came the big, larger-than-a-crouched-tiger metal boxes with what seemed like endless wires and knobs. To record aftershock sequences, portable seismometers were

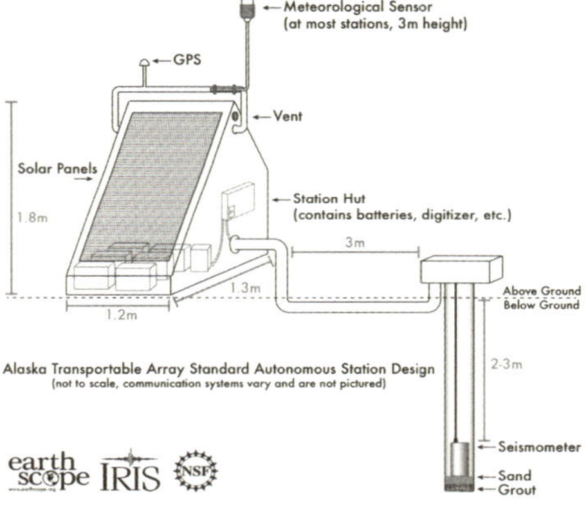

Figure. Schematic of an Alaska USArray seismic station. Image from Bob Busby.

designed using smoked paper. These seismometers were the size of an Xbox, and the scientist needed to carefully create a sooty smoke layer on the paper (avoid fingerprints!), while the device used a needle to etch (think record player) a non-soot seismogram trail. To set the recording, the smoked paper was dipped in shellac varnish, and these paper records were hung out to dry for a few days, just like you might air-dry your wash using a clothes line. Eventually, big became small, and strong motion seismometers were designed to be so sleek they looked more like really thick credit cards than recording instruments. These could be deployed anywhere, plugging into computers using a simple USB connection then—and this is my favorite part—duct-taped to secure the device.

The seismometers used in the NSF-funded USArray project are elaborate and are better explained with a schematic and photo.

Figure. Photo of an Alaska USArray seismic station. Photo from Bob Busby.

SH   I assume the primary events you measure are earthquakes, but what other phenomena or activity can seismometers detect?

DK   Every wiggle counts. A seismometer measures all ground motions close to where the seismometer is located. For the more expensive instruments that measure a wide range of frequencies, a seismometer can record signals of, for example, car doors slamming, traffic patterns in the region, and blustery wind days. In one project I worked on we found signals that were active for 20 hours, followed by a four-hour break. We deduced those signals were from cows milling about near the seismometer. Cows, we learned, sleep only four hours a day. The tricky part is untangling the non-seismic signals from the seismic signals, and the fun part is when a signal you thought was annoying noise becomes an important scientific finding.

An example of a "wow look at that" moment is when seismologists discovered what is now termed "seismic tremor." Seismic tremor is basically when the fault shimmies-and-shakes like your teeth on a cold day. Shake, rattle, and roll. Actually, it is just shake and rattle, no roll.

SH   Tell me about the USArray, the continental-scale seismic observatory. What goes into one of these monitoring stations and why do we need so many of them working in tandem?

DK   Scientists love data; I think that is true of any scientific discipline. If you break it down, science at its core is comprised of scientists making up all kinds of wild ideas (but they call them hypotheses because it sounds fancy) and then using data to prove if they are right or wrong. If you only have a little bit of data, it is harder to prove you are right. Scientists enjoy being right, and a proud scientist is a marvel to see (think NASA scientists and the Mars rovers). Moving our science forward by proving our hypotheses are correct is why data, lots of data, lots of good data, is important to scientists.

The USArray network is a network of ~400 seismic stations. The array was first deployed in 2004 in the western part of the US and iteratively rolled eastward until eventually in 2013 the USArray network had crawled its way across the full continental US. The USArray network was moved north to complete a similar spatial crawl across Alaska starting in 2014.

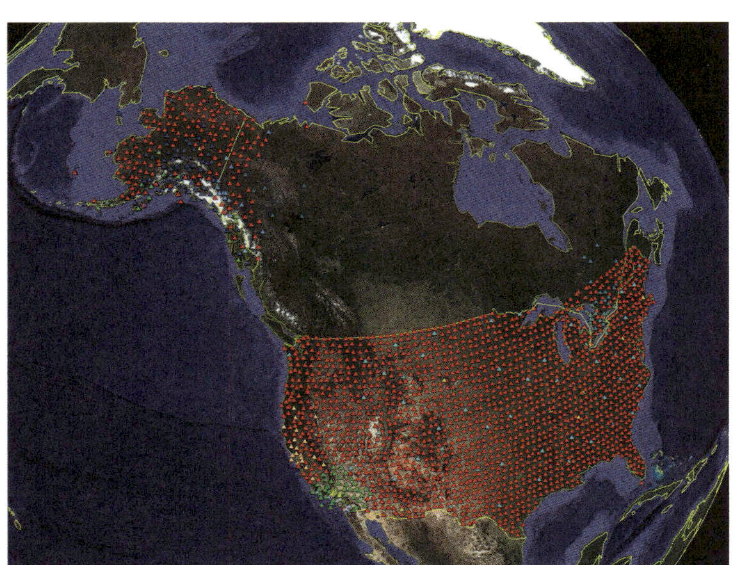

Figure. USArray station deployment (triangles; red operated by the Transportable Array; other colors from various contributed regional networks) map covering the lower 48 states and Alaska. Not all stations were operational simultaneously. The suite of ~400 stations were first deployed in the lower 48 traversing from the west coast to the east coast from 2004-2013 and then moved to Alaska in 2014. Image from Frank Vernon.

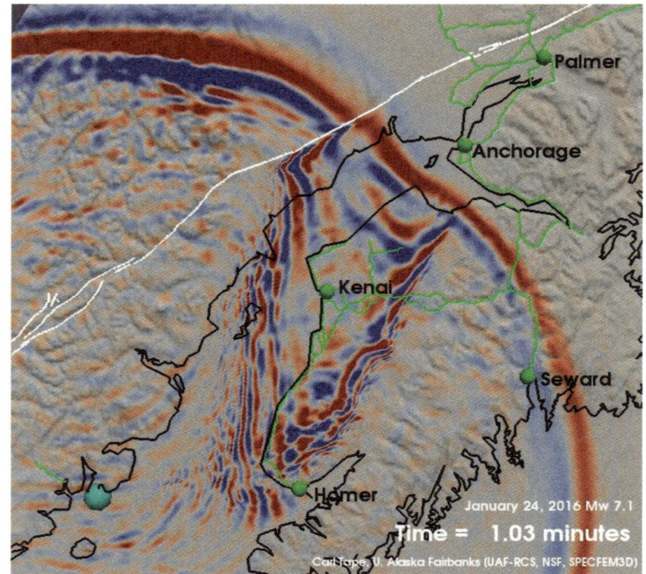

Figure. Example of how a basin structure can trap seismic waves. This is a computer model of the 24 January 2016 Iniskin, Alaska, magnitude 7.1 earthquake located to the southwest of this figure. The colors represent up- and down-motion, red and blue respectively. The initial seismic waves radiate outward from the earthquake in a circular pattern much like a water ripples outward from a stone dropped in water. We can see this behavior in the wide circular swaths of red and blue in the leading front of the seismic waves. Trailing in the wake of this circular front, is a distributed chaotic patch where the seismic waves are reverberating in the Kenai basin structure. Image from Carl Tape.

Importantly, the USArray stations were deployed in a grid pattern with inter-station spacing of ~70 km (~45 miles). The reason for a uniform grid spacing is so we can uniformly "extrapolate." Here, extrapolate means that we take a measurement at one location and assume the same measurement can represent what might be measured at other locations nearby. When you extrapolate, you have to guess what the approximate distance limit is where the extrapolation is valid. For seismic recordings, extrapolating out to two-giant steps is likely correct, but extrapolating from San Diego to Chicago is absolutely wrong.

I like the "working in tandem" part of your question; tandem spatial seismic recordings are key. Within our grid of seismic stations and tandem recordings, we can better discern what regions behave similarly and what regions do not. Some obvious differences emerge, such as "basin effects" that are caused by a basin of unconsolidated material. When seismic waves traverse a basin structure they often cause the material to shake like a bowl of jelly. The seismic energy gets trapped within the basin, and the resonance often greatly amplifies the seismic waves.

SH   You had the unfortunate [laughs] experience of trying to teach me the basics of Matlab. So, let's talk about SeisSound, the Matlab script you created for producing audible sound files and spectrograms from seismic activity. What was the original intent for SeisSound?

DK   Turning seismic data into sound is not a new concept; it has been around for almost seven-decades, since ~1953. My friend and colleague Zhigang Peng wrote the original Matlab code. My motivation for updating Zhigang's code to develop an IRIS data product was financial. Because I play at the intersection of disciplines, not fitting a norm, in my current position I need to raise 93% of my funding. This has pros and cons. The main con, of course, is constantly running on the hamster-wheel of collecting funding, but the pro is that I only work on projects that I love.

To me, the original intent for SeisSound was to better understand the frequency content within a seismogram and to better understand what one can and cannot record. Let's break it down. If you look at the seismogram of an earthquake in the time domain, you are looking at the temporal evolution of the ground motion amplitudes; larger ground motions are recorded as larger amplitude signals. In fact, some of the larger amplitude signals might overprint the smaller amplitude signals, making them impossible to discern. To gain a better glimpse of what we might be missing in the time-domain signals, we use frequency domain depictions, images called spectrograms that display not only amplitude as a function of time but also the frequency content. This turns a two-dimensional depiction of data (x-axis and y-axis information) into a three-dimensional depiction (x-axis, y-axis, and color-coding). For most, it is easier to understand the concept of low and high frequencies when you hear the sounds. High frequencies are squeaky and high-pitched, and low frequencies are lower rumblings. James Earl Jones's voice is primarily low frequency, whereas SpongeBob SquarePants' voice is primarily high frequency. Let's say James Earl Jones and SpongeBob are singing a duet. Most people will likely hear both the low-frequency and high-frequency sounds, and, after being trained, they will be able pick them out on a spectrogram. But from a time-domain recording of the sounds, you can't always discern all the intricate differences.

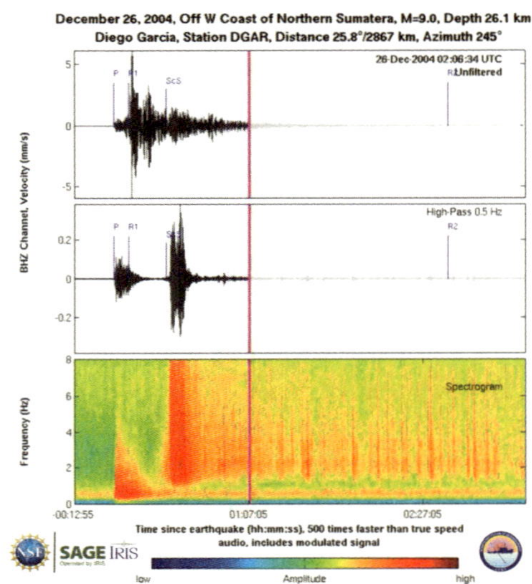

Figure. Seismic data from the 26 December 2004 Northern Sumatra magnitude 9 earthquake recorded at station Diego Garcia (station code DGAR) located on the Chagos Islands, in the Indian Ocean. These seismic data are displayed in the time domain (top) unfiltered data and (bottom) filtered data, high passed at 0.5 H. Select seismic phases are indicated, showing that as expected the P-wave is the first seismic phase to arrive at the station. Image from IRIS.

Another intent for SeisSound is to more easily identify aftershocks, either local aftershocks (i.e., those located near the mainshock, within one- or two-mainshock fault lengths) or distant aftershocks (i.e., 10's of mainshock fault lengths away). In a seminal paper, my all-time seismology hero Dave Hill identified aftershocks ~775 miles away from the 1992 magnitude 7.4 Landers earthquake. Amazing. Prior to Dave's study, the search for aftershocks from a magnitude 7.4 earthquake would encompass, at most, a ~100 mile radius. Dave's findings fundamentally changed our understanding of the physics of aftershocks. My favorite part of this story is that Dave's results even passed my beloved statistical test, one I fondly call the "well duh" test where statistics are not really needed because the conclusions are so obvious. I'll often re-read Dave's study as it inspires me to take time to look for the odd outlier and to not dismiss any anomalous findings. Spurred by Dave's work, numerous other remote aftershock sequences have been identified. One sequence that also passed the "well duh" test was the 2002 magnitude 7.9 earthquake in Denali,

Alaska, which triggered a series of aftershocks along the western coast of the US. A study by Pankow et al., in 2004, reported a 10-fold increase in seismic activity in Utah in the first 24 hours after the Denali mainshock.

The spectrograms in the SeisSound data products can help identify small remote aftershock signals, which are hard to pick out in time-domain data alone. The frequency domain representation of these remote aftershocks is easier to identify because these smaller earthquakes contain higher-frequencies that prominently stand out within spectrograms of large mainshocks.

SH   You've inhabited the intersection of science and art for a couple of decades now, from your work with scientific visualization, to the design of educational video games, to the win-win of integrating graduate student instructors into K-12 education programs. What's next, or put another way, what's been left out?

DK   It's complex, but, to boil it down into simple terms, this is how the intersection of science and art played out for me over the years. I was on the science side of the vast divide between science and art, and fortunately those before me started building a bridge between these two disciplines. This allowed me to see new connections and venture out onto that bridge, at first with trepidation, then with genuine curiosity, and now with a purpose to engage and interact. But I've always come to that bridge of engagement from the science side, peppering art into the science world. What I'm most interested in doing next is starting on the art side (a community I don't belong to) and peppering in the science.

SH   OK, let's say you get your wish, what science would you bring to the art world?

DK   A "draw an earthquake spectrogram" hands-on activity configurable for K-6th graders who have an interest in art. Art media? Crayons on parchment paper so that as the art is created the student can hear the evolution via the crinkling of the parchment paper. Scribbling? Encouraged. Factual? It's art, you decide. Goal #1 is drawing an earthquake using crayons — no rules, just go for it. Goal #2 is drawing an earthquake spectrogram. To ramp up for this activity, students will have the option to "dance a seismogram" to learn about amplitude and frequency and get our wiggles out (pun intended). An important component to solidify the student's understanding of amplitude and frequency will be listening to the seismic sound snippets from your project. For the art project, students are encouraged to use warmer color crayons (reds, oranges) to represent large amplitudes and cooler colors (blues, greens) used to represent smaller amplitudes. Low-frequency energy is going to be on the bottom part of the drawing and higher frequency energy toward the top. Time will evolve from left to right on the paper, similar to how we read a book.

A single earthquake usually consists of three phases named the P-wave, S-wave, and surface-wave. The P-wave is always the fastest of the three. Always. In second place is the S-wave and third the surface waves. The P-wave typically has the lowest amplitudes, so students will use green and blue crayons to draw their earthquake P-wave. The S-wave often, but not always, has larger amplitudes than the P-wave (i.e., use more orange and red colors). And typically, the star of the show is the surface waves that have much larger amplitudes than the P- and S-waves, so the student's red and orange crayon's will primarily be used to draw their surface waves, which will also last longer than the P- and S-waves.

Students will be encouraged to include a legend that indicates what the different colors represent. In this way, if they don't like the blue/green/orange/red color palette, a new palette can be designed. Scientists love legends, but scientific legends are typically bland and dull, aiming only to be informative. To spice things up, we might explore designing poetry legends in the form of haiku, rhymes, or even pure nonsense. It's art, not science, so creativity is at the forefront of the activity.

SH   I am also venturing out onto the connecting
bridge, but I'm always coming from the art side. I recently
completed a music project using the ultrasonic echolocations
of bats as compositional course material. Is this project the
inverse: bringing infrasonic frequencies into the realm of
human hearing? Is this the Earth's voice speaking if we only
had ears that could hear it?

> DK   You are describing the sweet spot between art and
> science — the teeter-totter of play. I like the idea of the
> Earth having a voice, and it is our job to listen to what
> is said. In 1969, four seismometers were installed on the
> moon by the Apollo mission, allowing us to record and
> listen to the moon's voice too. One thing we learned was
> that moonquakes are much deeper than earthquakes.
> More recently, scientists recorded the voice of Mars. At a
> seismology meeting in 2019, I sat in rapture as the key-note
> speaker unveiled data from a marsquake — an amazing hush
> fell over the room. I'm sure scientists will be marveling and
> studying the voices of our Earth and planets for years to
> come and that what we hear and don't hear will evolve with
> time.

SH   I love the combination of incredibly sophisticated
technology with what seems to be a pretty crude method of
delivery: basically a steel box and auger hole, hand drilled by
a couple of folks. I want readers to remember that science is
both a Matlab script and a rusty pickaxe. Can you talk about
how these two parts of the puzzle work together, not just in
geophysics but in the wider realm of science?

> DK   As you suggest, how well we perform our science
> often boils down to what is in our tool kit, how we can
> improve and use that tool kit, and how we can team up
> with others who have complementary tools. I think of it as
> a spiral. At the base of the spiral, within your very narrow
> discipline, you select only a few similar-minded scientists to
> work with and learn from. This is a comfort zone that many
> never escape from. But for most, the circle expands from,
> for example, a seismologist working with a geomorphologist
> or with a glaciologist, and, when big-data comes into play,

a computer scientist. Surprisingly, although it shouldn't be, as you move out along the spiral, the discourse between different disciplines becomes challenging. A word that we scientists think we know means different things within the different disciplines; frustration seeps in, and the easy route is to retreat back into the inner spiral. The more we can move science from the innermost spiral to the outer edges of the spiral that combine multiple disciplines, to my mind, the more we will learn.

SH   I know you studied music before you became a seismologist. Did you think these two disciplines could overlap such that you'd be asked to write about the musical potential of seismic activity?

DK   Absolutely! I love working at the intersections of disciplines, especially within the explosion that can happen when you merge art and science. I'm happy our nation is moving toward intertwining art and science in schools in what has been coined STEAM (science, technology, engineering, art, and math).

This is what I heard when I was growing up. Teacher: What do you want to be when you grow up? Kid: A teacher or a fireman or a movie star (i.e., the few jobs they knew about). As the kid grew up, the answers sometimes morphed to include a lawyer or a doctor, perhaps driven by what some grown-ups wanted to hear. Full disclosure, my twin-sister is a lawyer, so I'm not dissing lawyers and doctors; the point is that there was a limited selection of this-is-your-life discrete career tracks, and kids were often asked to pick one and jump on at an early age. Typically, there were no tracks for art/science interdisciplinary work. If you tried to jump back-and-forth from the science track to the art track, SPLAT you get jettisoned into the mush pot.

Pairing art and science is an essential part of my world. I hope more resources are put into promoting these types of non-standard fields of study — especially promoting STEAM ideas to our younger generation. I expect there are many kids who are angst-ridden because they are not able to answer "what do you want to be when you grow up?" and they drift from thing to thing. A shift is needed to ask questions more along the line of "what are you most curious about?" If the items of curiosity span multiple disciplines, then bingo, we have found an interdisciplinary soul.

Acknowledgments

Deep gratitude for our generous funders and visionaries:
The Anchorage Museum
The National Geographic Society

This project only happened because of the consistent
support and insight provided by the amazing Hollis Mickey.
Huge thanks to Julie Decker, Ryan Kenny, Aaron Leggett,
and the entire exhibition staff at the Anchorage Museum.

Also, thank you:
Dr. Maria Williams for sharing your passion and knowledge
of Alaska Native Indigenous cultural practices.
The Switzer-Strauss family for your hospitality.
Ryan Bierma, Bob Busby, Jeremy Miner, Perle Dorr,
Manoch Bahavar, and Chad Trabant at the IRIS Consortium.
Sarah Urist Green for your much needed input and critique.
Jeremy deVine for believing in the musical experiment
called *Field Works*.
Eleonoor Jap Sam for making this book possible.

Most of all, thank you to the contributing musicians,
essayists, poets, and scientists whose vibrant new works
make these pages come alive.

Contributors

BOOK

Anaïs Duplan is the author of *Blackspace: On the Poetics of an Afrofuture* (Black Ocean, 2020).

Ashon T. Crawley is an interdisciplinary performance and visual artist, writer, and teacher based in Charlottesville, Virginia. He is author of *Blackpentecostal Breath: The Aesthetics of Possibility* (Fordham University Press) and *The Lonely Letters* (Duke University Press). All his work is about otherwise possibility.

Debi Kilb is a seismologist at Scripps Institution of Oceanography.

Enrique Ramirez is a writer and a historian of art and architecture. His work considers histories of buildings, cities, and landscapes alongside larger cultures of textual and literary production in Europe and the Americas from the Renaissance onward. He is currently a faculty member at Yale School of Art, where he teaches seminars on print history at the Beinecke Rare Book & Manuscript Library.

Gustavo Valdivia is an environmental anthropologist, sound artist, and documentary filmmaker. He has been a chapter scientist in the Intergovernmental Panel on Climate Change's (IPCC) Sixth Assessment Report (AR6), an artist in residence at Praksis Oslo, and a field producer for the documentary BBC series *Frozen Planet II*.

Hanna Benn is a composer, vocalist, and genre-spanning collaborator. Her multi-disciplinary approach has incorporated dance, opera, and theater—breaking boundaries and discovering new sonic landscapes in the process.

Hanna Craig is an Alaska-based filmmaker and multimedia producer specializing in documentary storytelling on themes of the environment, education, culture, and the arts.

Ishmael Angaluuk Hope explores and illuminates Indigenous sovereignty through his art and Indigenous critical theory. He lives, creates literary art, and raises a family of five children with his wife Lily Hope, a Tlingit weaver, in Dzantik'ihéeni, or Juneau, Alaska.

Joan Naviyuk Kane teaches at Harvard and Tufts, and was founding faculty of the graduate creative writing program at the Institute of American Indian Arts. In 2021, *Hyperboreal* was published in French translation by Editions Caractères and *Dark Traffic*, a collection of new poems, was published in the Pitt Poetry Series.

Kashina's ongoing scent journal started in 2017 and appears on Instagram @illicium.verum.nova. Her writing has been featured in *Affidavit*, *GARAGE*, and *Canadian Art*. Her collection *East of Borneo* was published by Antifurniture in September 2020, and she is currently working on a second book.

Laraaji is a multi-instrumental musician/recording artist
and laughter release meditation guide. He has recorded and
toured the globe with amazing ambient music performances
and therapeutic Laughter Meditation workshops since the
1980s.

Paige Lewis is the author of *Space Struck*. Their poems have
appeared in *Poetry*, *American Poetry Review*, *Ploughshares*,
The Georgia Review, and elsewhere. Lewis curates the
YouTube series Ours Poetica. They currently teach at Purdue
University and in the low-residency MFA program at
Randolph College.

ALBUM

AFRODEUTSCHE (Henrietta Smith-Rolla), is a British
born Ghanaian/Russian/German artist, BAFTA-nominated
composer, producer, and DJ based in Manchester, UK .

Ben Chatwin is a composer and producer working from
The Vennel Studio in Queensferry, Scotland. He creates
a musical language permeated with a tension between the
electronic and acoustic worlds.

Brad Weber is a drummer/producer/synth nerd from
Toronto. He makes music as Pick a Piper, Coy Haste and
also plays drums for the Caribou live band.

Carsten Nicolai is an artist and musician who lives
and works in Berlin. For his musical outputs he uses the
pseudonym Alva Noto. Inspired by scientific reference
systems, Nicolai explores mathematical patterns such as
grids and codes, error and random structures, as well as the
phenomenon of self-organization.

Deantoni Parks is a creative technologist from Newnan, GA. He is a member of Jean-Michel Basquiat's Industrial Art Noise Band, Gray and the founder of Technoself School of Philosophy.

Green-House is the project of Los Angeles based musician Olive Ardizoni. Their music is written with the intention to enhance living spaces for humans and non-humans alike.

Hanna Benn is a composer, vocalist, and genre-spanning collaborator. Her multi-disciplinary approach has incorporated dance, opera, and theater — breaking boundaries and discovering new sonic landscapes in the process.

Janie Cowan is an internationally touring double bassist from Anchorage, Alaska. She now works out of her experimental studio in the Mojave desert, recording and hosting groups from deep and diverse cultural legacies.

Laraaji is a multi-instrumental musician/recording artist and laughter release meditation guide. He has recorded and toured the globe with amazing ambient music performances and therapeutic Laughter Meditation workshops since the 1980s.

Masayoshi Fujita (aka el fog) is a Japanese vibraphone player and drummer based in Hyogo. Using experimental techniques, he fuses jazz, ambient and electronic sounds and has collaborated with Jan Jelinek and Guy Andrews.

Nathan Fake is a musician and performing artist from Norfolk, UK. He recently released his fifth studio album, *Blizzards*.

Olga Wojciechowska is a composer and sound designer based in Poland. Much of her personal work was released anonymously under the pseudonym Strië.

Penelope Trappes is a UK-based, Australian-born vocalist, musician, and ethereal soundscaper signed to Houndstooth.

Qasim Naqvi is an electronic musician and drummer based in Brooklyn NY. He is a founding member of the group Dawn of Midi.

Randall Taylor is an audio/visual artist and musician based out of Portland, Oregon. He performs and creates music under the name Amulets and has been outputting tape based music and experiments since 2011.

Sophia Loizou is an artist, music producer, and academic from Bristol, UK. She lectures at the British & Irish Institute of Modern Music and is a doctoral researcher at Goldsmiths, University of London.

EDITORS

PRINTtEXT is a creative studio and collaborative publisher founded by W Benjamin Blevins and Janneane Blevins.

Stuart Hyatt is an artist and musician from Indianapolis. He is a National Geographic Explorer and the creator of Field Works.

Colophon

PUBLISHER
Jap Sam Books, Prinsenbeek, the Netherlands
www.japsambooks.nl

PRINTING & BINDING
Graphius, Ghent, Belgium

Typeset in GT Flexa designed by Dominik Huber with Marc Kappeler from Grilli Type; Adobe Caslon Pro designed by Carol Twombly for Adobe Orginals; and Interstate designed by Tobias Frere-Jones for Frere-Jones Type.

Cover printed on 120 g/sm² Wibalin Natural Pumice with holographic foil stamp. Interior printed on 120 g/sm² Munken Polar Rough White.

ISBN 978-94-92852-35-9

EDITORS
Stuart Hyatt, Janneane Blevins, and W Benjamin Blevins

DESIGN
Book and cover design by PRINTtEXT / Janneane Blevins and W Benjamin Blevins
www.printtext.co

PHOTOGRAPHY
Unless otherwise noted, all images courtesy of The IRIS Consortium.

CONTRIBUTORS
Hanna Benn, Hanna Craig, Ashon Crawley, Anaïs Duplan, Ishmael Angaluuk Hope, Joan Naviyuk Kane, Kashina, Debi Kilb, Laraaji, Thomas Lecocq, et al., Paige Lewis, Enrique Ramirez, and Gustavo Valdivia

Proofreading by Aaron Bogart, Eleonoor Jap Sam